THROWAWAY NATION

THROWAWAY NATION

The Ugly Truth about American Garbage

Jeff Dondero

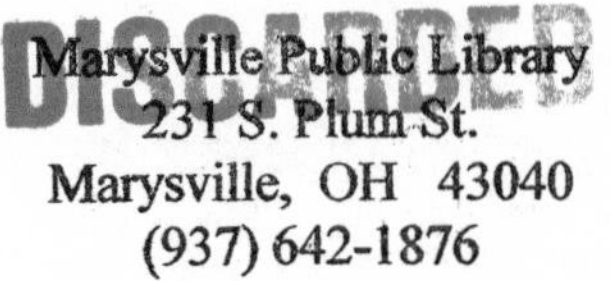

ROWMAN & LITTLEFIELD
Lanham • Boulder • New York • London

Published by Rowman & Littlefield
An imprint of The Rowman & Littlefield Publishing Group, Inc.
4501 Forbes Boulevard, Suite 200, Lanham, Maryland 20706
www.rowman.com

Unit A, Whitacre Mews, 26-34 Stannary Street, London SE11 4AB

British Library Cataloguing in Publication Information Available

Library of Congress Cataloging-in-Publication Data

Names: Dondero, Jeff, 1947–
Title: Throwaway nation : the ugly truth about American garbage / Jeff Dondero.
Description: Lanham, Maryland : Rowman & Littlefield, [2019], an imprint of The Rowman & Littlefield Publishing Group, Inc., | Includes bibliographical references and index.
Identifiers: LCCN 2018029445 (print) | LCCN 2018044311 (ebook) | ISBN 9781538110331 (electronic) | ISBN 9781538110324 (cloth : alk. paper)
Subjects: LCSH: Refuse and refuse disposal—Social aspects—United States. | Salvage (Waste, etc.)—United States.
Classification: LCC HD4482 (ebook) | LCC HD4482 .D66 2019 (print) | DDC 363.72/88—dc23
LC record available at https://lccn.loc.gov/2018029445

∞™ The paper used in this publication meets the minimum requirements of American National Standard for Information Sciences Permanence of Paper for Printed Library Materials, ANSI/NISO Z39.48-1992.

Printed in the United States of America

CONTENTS

ACKNOWLEDGMENTS AND DISCLAIMER

ACKNOWLEDGMENTS

The author wishes to thank the many people who assisted in the creation of this book, whom I call curators of information. These are the people who tirelessly gather facts and figures to illuminate the world and to give others something to ponder.

As always, very special thanks to my ad hoc editor and sounding board, Patrick Totty, for advice, confidence, encouragement, suggestions, conversation, a few cocktails, and aid in many ways.

To my editor, Suzanne Staszak-Silva, with unflagging patience in getting this large undertaking in order. To my production editor, Andrew Yoder, and to Rowman & Littlefield for giving me the opportunity to allow others to read what I write.

DISCLAIMER

It is important to remember, "There are lies, damn lies, and statistics." Most of the stats here have been gathered by local, state, federal, and worldwide government agencies; independent scientists and writers; universities; various water and power companies; and other informed sources. Consequently, I do not claim that all the statistics presented represent accurate and true statements, percentages, and facts, and I do not warrant or make any representations as to the content, accuracy, or completeness of the information, text, graphics, charts, web links, websites, and other items contained in their media presentations.

Aggregating and writing information for this kind of book has its inherent problems and predicaments. When presented with questions, people use different ways to find diverse answers and conclusions. Consequently they may vary, sometimes quite a bit. In addition, some of the facts presented might be affected by time, changing world events, or new discoveries. As in many things, opinions vary as to number, percentages, predictions, and the veracity or divergence of the results obtained by individuals using the same information. No one can completely and accurately predict the future. And not all scientists would agree on matters such as global warming, climate change, or that CO_2 is contributing to the detriment of our environment.

What I have tried to do is present an informed opinion about what the treatment of our planet, its resources, and our trash might hold for the future of the rock in space upon which we live. I have also stated that there have been doomsayers who make dire predictions about the fate of the Earth. I hope I'm not one of those people, but I figure it's best to err on the side of caution and not carelessness, while pointing out that I have unflagging confidence in the human race to confront and solve the problems of the day—or century.

Although most of the facts presented herein are defensible, I use them as literary, entertainment, and educational devices to give the reader a general perspective on the subject of the resources, energy, and the waste we create. In an effort to communicate more easily and effectively, I have taken some averages, mean numbers, and common sense, and I have modified statements to reflect more than one set of opinions and/or updated them using logic and informed opinion.

The lawyers put it another way: Neither the publisher nor any of its employees makes any warranty or guarantee, expressed or implied, or assumes any legal liability or responsibility for the accuracy, completeness, or usefulness of any information, percentage, apparatus, product, device, or process disclosed, or represents that its use would not infringe upon privately owned rights. Reference herein to any specific commercial copyright, product, process, or service by trade name, trademark, manufacturer, or otherwise, does not necessarily constitute or imply its endorsement or recommendation.

The views and opinions of authors expressed herein do not necessarily state or reflect those of the Rowman & Littlefield Publishing Group or its agents. The information and/or products mentioned herein do not imply an endorsement, guarantee, or warrantee. This book is for entertainment and educational purposes only.

PREFACE

Everyone Wants a Pet, but No One Wants to Pick up the Poop

If we were guests, we would have been asked to leave. As proprietors, our property value would have plummeted. As groundskeepers, we would have been fired. Just because we are the dominant species doesn't mean we own the place and can be passive guests. We have the responsibility to leave it for those who come next. We have influenced conditions on this planet throughout its history—and objectively not for the better. It seems that it wasn't enough to befoul just the planet; now we are also leaving our left-behinds in space and on other planets.

We are indebted to earlier life forms for contributions that range from the oxygen in the air we breathe to fossil fuels upon which our civilization has become dependent. We've repaid that with pollution. It is unlikely that the biosphere as a whole will be endangered by our actions; it has survived bigger calamities in the past. However, humans and other livings things might not fare so well. Species are dying out at the fastest rate since the demise of the dinosaurs, 65 million years ago. Ninety-nine percent of all living things on Earth have died before us. Unlike the previous five mass extinctions, which were mostly caused by monumental natural disasters, the next event will be entirely due to human impact. And hopefully, we are not staring into the sixth, and next, extinction event.

That is scant comfort, because we are vulnerable to even modest climate changes that persist only for decades or centuries. We *Homo sapiens* can ill afford something that seems as trivial as a few degrees increase in the temperature of the Earth or a boost in the frequency of natural disasters or even modest sea-level rise. Our planet may seem robust—after all, life has been on the planet for more than three billion years—but it can nonetheless appear fragile from the perspective of individual species, especially us. That is why there is cause for concern about the global environmental consequences of our personal, agricultural, and industrial practices. The matter has generated lively debates, but they often end in stalemate, as some insist that humanity will endure any catastrophe, while others contend that we are under considerable stress and might not survive—except a few lucky ones that flee to another planet, hopefully not to start dystopian practices all over again.

Climate change is a normal part of the Earth's cycle, a natural variability related to interactions among the atmosphere and gases, ocean, land, and changes in the amount of solar radiation reaching the Earth. The geologic record includes a plethora of evidence for large-scale climate changes over all scales of time. The problem we face now is the minute amount of time it has taken for alarming changes.

We think of all the problems of the biosphere—the surface, atmosphere, and hydrosphere of the Earth occupied by living organisms—as separate and we try to deal with them one at a time. But Earth's ecosystems are not separate; they are interlinked by the ebb and flow of energy, fresh and salt water, nutrients, land, air, plants, and animals into one fabric, along with the use and abuse of the Earth's resources and waste.

Of all the known living species on Earth (about 8.7 million give or take, with 6.5 million on land and 2.2 million in oceans) humans alone can produce waste that nature cannot biodegrade, can destroy its own habitat, and can cause the extinction of others. And according to the Census of Marine Life, less than 15 percent of species have been discovered—that leaves 86 percent that are still unknown to us, or about another 130.5 million species. Because we are changing some of our land mass and the oceans, many species will become extinct before they are discovered, as surely some already have.

A spill from a pesticide plant turned the Rhine into a "dead river" the entire length of Germany; acid rain that falls in Scandinavia kills fish; CO_2 and oil spills destroy the coral reefs; air pollution kills vegetation in the Los Angeles basin; radioactive waste dumped into the Irish Sea can be traced up to the Arctic Circle and over to Denmark; and virgin forests in the Amazon basin are clear-cut and lost forever.

Even the most tightly woven fabric will fall apart if it has enough holes poked in it. No one knows how many such holes our fabric can sustain. And some human activities, such as fossil-fuel burning, are now so pervasive that they are changing the atmosphere and the climate.

My generation, the Baby Boomers, are paradoxical as well as poignant about the world in which we live. We are the only generation that simultaneously fought for a greener world and at the same time created a throwaway culture capable of burying itself in its own garbage. Now it is our liability to try to turn the tide of trash and find ways to avoid the tilt toward the devastation of most of the life forms on Earth. A teeter-totter timetable is tilting for the first time in human history toward something that we caused but might not be able to contain or change. It began with the Industrial Revolution and goods being produced in mass quantities; became exacerbated after World War II, especially with the expansive production and use of petroleum and its most ubiquitous offspring, plastics; and exploded in the logarithmic increase of toxic e-waste. All this has made us the ultimate throwaway nation, one that is proportionately small in population but enormous in terms of resource use, waste produced, and damage done.

"There has never been a time of greater promise or greater peril," says Professor Klaus Schwab, founder and executive chairman of the World Economic Forum. This might even be echoed by the Bible, "And I brought you into a plentiful land, to eat the fruit thereof, and the goodness thereof, but when ye entered, ye defiled my land, and made mine heritage an abomination" (Jeremiah 2:7).

Before Steven Hawking died, he said that the best bet for humankind was to leave the Earth. Others have hope and think there is still time to clean up our home. But we have to have fewer children, husband our resources, and stop crapping in our own nest if we want our progeny to stick around and live on a healthy planet Earth. And it all starts with every single one of us not only doing our best to conserve and take ownership of our trash, but also putting pressure on everyone else, including politicians and big business.

I have faith in our ability to avoid an environmental Armageddon because although humans are sometimes careless and inattentive, we are also amazingly ingenious and resilient. It is my hope that this book will not only provide some "ah-ha" moments, but also generate hope, enthusiasm, answers, and a commitment to the belief that the human mind and spirit will conquer the problems that we face now and that await us in the future.

1

TALKIN' TRASH

The greatest threat to our planet is the belief that someone else will save it.
—*Robert Swann*

A LITTER BACKGROUND

America used to have a litany about waste. It was forged and passed down from our Puritan and Calvinist heritage, case-hardened via the Great Depression and a world war where almost nothing was wasted, to a generation that has broken that ancient legacy with a new mantra—buy new, trash the old. The buy by-product—the pollution of our planet.

The day of repairing and repurposing things is almost a lost art. Today, if it's reached its due date of planned obsolescence, if it's passé or broken, we toss it. Cynics call it modern marketing, but there is also a seductive ease in leaving things behind that don't work or are not fashionable—forget it and get something brand-new, bleeding edge, and chichi.

It's high irony that the generation that fashioned ecology, reenergized recycling, husbanded alternative energy, made "living less and green" fashionable, now has more of everything than any other generation and is also the age group that is burying the planet and outer space around it with their spoils. We have become a throwaway nation.

With the advent of plastic pouches, spray cans, and squeezable throwaway tubes used for everything from food to cosmetics came a new convenience of disposability to everyone's lifestyle. Easily disposable items offered a new freedom that was quickly linked to the notion of abundance, and it leaked from the home and the workplace to the world outside.

Four complete planets the size of the Earth would be required if undeveloped countries consumed at the same rate as the United States. And each U.S. citizen produces about 1,606 pounds of garbage a year. About 55 percent of all that trash is buried in landfills. What's alarming is that number is twice what it was in 1960. One landfill site in Virginia alone is comparable to 1,000 football fields in length and the height of the Washington Monument—555 feet.

Landfill trash does not decompose as, say, compost does. It is sometimes buried with very little regard to composition and with little oxygen or moisture surrounding it, which inhibits decomposition. Nonbiodegradable items such as plastic bags and diapers can remain buried in the Earth's landfills for hundreds of years in a state similar to when first buried, while creating very noxious greenhouse gases. Someone once suggested that the fine for littering should be being made to do time in a dump.

MOVABLE DUMPS

In 1987, an orphaned garbage barge chugged up and down the East Coast looking for a place to dump its rank cargo. The *Mobro* was a poster child for an unwanted load of rubbish looking for a landfill in which to unload its sordid freight. Originally scheduled for dumping in North Carolina, the *Mobro* carried six million pounds of pungent New York garbage, but it got turned away from its destination and was even threatened with the state National Guard if it tried to unload in North Carolina. It tried to take a dump in six other locations and was turned away from all.

It sparked a firestorm of "Is there nowhere to throw away our throwaways?" Finally, the barge wound up getting permission to dump the controversial cargo back in Islip, New York. But before that happened, a judge ordered the load to be burned in Brooklyn.

Today, 23 tons of garbage leave New York every day; 2,600 *Mobro*-sized barges travel a half million miles a year to more than 600 methane gas projects (a by-product of garbage and landfills) that produce 15 billion kilowatts (kW) of energy annually. A gigawatt is a billion watts and a gigawatt of power will provide enough energy for about 700,000 homes annually. By the way, it costs New York more than $241 a ton to have it hauled away and dumped.

THE LOGIC OF LITTERING

The reasons for littering are legion, starting simply with laziness. People come up with countless excuses to throw their trash on the ground instead of taking the short amount of time to find a proper garbage can. If it is more than 12 paces or approximately 10 to 12 seconds away from a garbage bin, chances are good that it will be littered.

Littering, however small, does have consequences, and here are some facts that may surprise you, including ecological, financial, legal, and even moral consequences. Litter can cause severe accidents. All that it takes is trash in the road, and it's an accident waiting to happen.

According to an article in the journal *Science*, litter and clutter contribute to discrimination between peoples. When encountering others in an unkempt or shabby environment the mind begins to sort and classify people, noting differences, increasing the chance of stereotyping and division. Scientists find people of different races sit further away from each other in dirty, disorderly environments.

A WASTE OF HISTORY

Waste and its problems are as old as humankind. Archeologists can figure out many things about a society or civilization by what it leaves behind. In North America, archaeological studies have found that Native Americans in what is now Colorado may have produced an average of 5.3 pounds of waste per person per day in approximately 6500 BC—one pound more than the average American today.

As early as 79 AD, some early recycling methods included feeding vegetable waste to livestock and using green waste as fertilizer. As late as 1800, many women carried small pouches of perfume or flower-scented potpourri to avoid fainting from the rotten "vapors" in cities.

By the mid-1700s, Americans began digging refuse pits in municipalities for their household wastes, rather than throwing the garbage into the streets and alleys. In 1739, Benjamin Franklin and his neighbors unsuccessfully petitioned the Pennsylvania Assembly to stop waste dumping and to remove tanneries from Philadelphia's commercial areas. Philadelphia instituted the first municipal street cleaning service in 1757.

In the 1900s, "piggers" were used, where hogs were fed garbage. Supposedly, 75 swine could consume a ton of refuse per day. In the 1920s, communities began to realize and plan some sort of organized citywide waste collection and disposal service, creating dumps or landfills and providing such services. However, many were located near rivers and streams, where foul liquids and refuse from the dumps could easily enter the water table and threaten clean water supplies. They were also extremely unsanitary, attracted vermin, gave off repugnant odors, and were even fire hazards.

It was not until 1929 that the federal government issued the first location restriction for disposal sites by recommending, but not requiring, dumps to be located away from river banks. In 1934, the United States Supreme Court upheld a lower-court ruling requiring New York City to cease disposal of its municipal waste at sea. In the 1930s, California passed laws prohibiting disposal of solid waste within 20 miles of shore.

Around 1934, the first container deposit laws were enacted by the National Recovery Administration requiring deposits of two cents for small bottles and five cents for large ones. Concern about litter arose again during the late 1960s and early 1970s, when one-way containers were used for about 60 percent of the volume of soda pop in the United States and about 75 percent of beer. During that time, more than 350 proposals were introduced in Congress, in state legislatures, and in local legislative bodies that included bans on nonrefillable containers.

Deposit and return was common and consumers saw getting their money back as a normal part of purchasing something in a bottle. But those empties were bedeviling the beverage industry by the sheer amount of buying, distributing, collecting, and returning bottles to the factory for cleaning. There was also the cost of reproducing heavy glass bottles, as well as the bottom-line hassle of stores "pawnbroking" bottle deposits.

In 1965, the federal government finally put the solid waste problem on a par with protection of water resources and passed the Solid Waste Disposal Act (SWDA), the federal government's first effort to implement a comprehensive management framework for the nation's solid waste. In 1970, Congress passed the Conservation and Resource Recovery Act (RCRA), shifting the emphasis of federal involvement from disposal to recycling, resource recovery, and conversion of waste to energy, and implementing a national system for hazardous waste management.

The RCRA, defines solid waste as any garbage or refuse; sludge from a wastewater treatment plant, water supply treatment plant, or air pollution control facility; and other discarded material resulting from industrial, commercial, mining, agricultural operations, and community activities.

Also, that year the Environmental Protection Agency (EPA) was created by President Nixon, and politicians of all political persuasions began to respond to the serious problems of waste in the country.

WE'RE NUMBER ONE

After World War II, the United States became the richest, most powerful, and prolific manufacturing country in the world. The years between 1946 and 1960 were the beginning of the "good life" for America as we witnessed an amazing expansion and consumption of goods and services. Gross national product (GNP) rose by 36 percent and personal consumption expenditures by 42 percent. And with that came and the Aladdin's Lamp—or was it the Pandora's Box—of waste and junk.

Detroit filled our desire for big, cushy, comfortable cars so we could travel on the new interstate highway system that was soon dotted with fast-food franchises. As the economy boomed, the convenience and variety of eating and drinking outside home and work blossomed and litter followed like a bad habit. When McDonald's remodeled and revamped their restaurants during the 1950s, they tossed out every dish and glass, metal spoon, knife, and fork, replacing them with paper and plastic—which contributed hundreds of tons every year to the municipal dumps and landfills—and to the countryside, as inconceivable amounts were simply just tossed out car windows.

Mountain ranges of debris filled America's landfills and went, for the most part, unnoticed and ignored. An unwished-for and unwanted high tide of trash followed—and the vultures finally came home to roost.

Perhaps because we have so much, we don't think about the amount that is grown and tossed, built and chucked aside, thrown away for the newest gadget—gorging on goods overproduced, underused, and unrecycled.

Although Americans recycle far more materials than we did 40 years ago, we also throw away much more than we did back in "the day." In 1960, our citizens disposed of 2.7 pounds of trash per person each day. Today we're up to 4.3 pounds and gaining weight for a total mass of more than a quarter of a billion tons each year.

TRASH TRADE

There are approximately 10,000 shuttered landfills in the United States, many of which are finding new life and repurpose as gas and material resource bonanzas. The trash trade is a $4 billion industry, and many state landfills are only too happy to take garbage from other states—for the right price.

Some scientists claim that dumps and landfills will be a place of recovery for various materials and a kind of archeologists' lab of the past for future fossil and sociological records.

Even though methane, produced both naturally and otherwise, doesn't linger as long as carbon dioxide, it's far more effective at absorbing the Sun's heat, thus contributing to global warming. For the first 20 years after it meets the atmosphere, methane is 30 times more potent as a greenhouse gas than carbon dioxide.

Although recycling has increased in recent years, so has trash generation. Six times as many PET (plastic) water bottles were thrown away in 2004 than in 1997, and 10 times more since 2004, or about 167 disposal water bottles per person each day in the United States. But only about 32 percent get recycled.

WHO INVENTED THE LITTERBUG AND WHY?

The derivation of the term *litterbug*, like many coined words, seems to be in both dispute and disrepute. Some claim it was derived by a Mrs. Hand from Florida around 1950 and used as a promotion for a roadside beautification campaign. Others claim that is was created by Paul B. Gioni, a copywriter in New York City who came up with the popular television campaign theme in 1963, "Every Litter Bit Hurts." The moniker was hyped by the "Keep America Beautiful" (KAB) organization as part of a campaign to vilify those who tossed trash in public places. It stuck in the American lexicon like bubblegum on the bottom of a shoe.

Technically, litter is rubbish as small as a gum wrapper or as large as a car that has been tossed aside irresponsibly. Litter can originate anywhere from moving vehicles to building sites. It can be costly if you get caught tossing it too. Legal fines can range up to $10,000 in some states. It causes damage to the flora and fauna in our living spaces. It looks lousy to everyone, and can lower property values.

Manufacturers, not wanting to deal with bottles began using disposables stamped with the words "No Deposit, No Return." That created a cascade of broken and castoff containers, quickly becoming a nuisance in public places, and communities considered bans on disposables.

The next step by the container companies (after righteously proclaiming that litter was a serious problem confronting the environment and their businesses) was to figure out how the used consumer goods could become either unfashionable or obsolete while dealing with public outcry, and their own bottom line.

Their answer was masterful. In 1953, to ensure better public relations, they introduced "nonrenewable" packaging products—cue the aluminum and steel cans, paper plates, and plastic utensils—that could be produced, trashed, and then reproduced again while shifting responsibility from the retailer or manufacturer to the community or consumer. Hello recycling. Their campaign also included throwing some shade on consumers for being "litterbugs," but not on big businesses, which took no responsibility for their products except to realize that refillable containers were a hassle and vastly less profitable.

And it worked. By the late 1950s, anti-litter ordinances were being passed in municipalities and statehouses across the country, while not a single restriction on packaging made it into law. The container industry's public relations people came up with campaigns with taglines of "Packages don't litter, people do," and "People start pollution. People can stop it," and made litterbug and littering a household word.

Enter the KAB campaign. The founding members: tobacco giant Philip Morris; beer baron Anheuser-Busch and soft drink big shots Coca-Cola and Pepsi created KAB to clean up America; at least that's what their press releases said. The organization was also developing a push-back against the environmental movement's increasing pressure on the industry to clean up its act.

And who can forget the KAB crying American Indian in the famous 1971 TV commercial featuring Iron Eyes Cody, mourning that "Pollution is a crying shame," in his "native land." This should be taken with a bag of litter, as old Iron Eyes had been speaking with a forked tongue, insisting until his death that he was descended from a Cherokee father and a Cree mother. Actually Espera Oscar DeCorti was an actor of Sicilian heritage. But the ad became one of the most memorable and successful campaigns in advertising history and was named one of the top 100 advertising campaigns of the 20th century by *Ad Age* magazine.

Everyone, especially the container industry, got an unexpected windfall because by pushing for recycling, it mobilized the nation to do a lot of free labor by getting rid of an unsightly mess, and helped the public buy into the recycling trend. KAB activism and agit-eco-thinking stormed into the 1960s as people became obsessed about being enviro-minded by becoming the trash police trying to save Ma Nature one can, bottle, and piece of paper at a time. Litter became a banner battle cry, almost eliminating the industrial culpability and the burdensome business of deposit and return.

Even though many states have initiated bottle bills, many have failed, due to strong opposition from the beverage and grocery industries that claim they are in the business of selling goods, not recycling or buying trash. The beverage industry has a history of fighting new bottle bills by tooth and bottle opener, seeing them as a "direct and politically motivated infringement on the free market and a threat to profits," as Matthew Gandy wrote in *Recycling and the Politics of Urban Waste*.

To be fair, KAB did focus on some of the vital issues of litter prevention, like waste reduction, recycling, community greening, and beautification by organizing volunteers to clean up litter and illegal dumpsites in their communities, as well as removing graffiti and planting trees, flowers, and community gardens. And in recent years the KAB network has grown into more than 1,000 affiliates and participating organizations, and is well-positioned to coordinate a national-level campaign with local involvement, and doing a pretty good job.

The catch-22 is that recycling will only do so much to limit the production of containers or trash. Even with our enlightened attitude toward recycling, we still have a long way to go, and it's only part of the solution to the trash generated by this country. The more reasonable answer (except for manufacturers) is to stop creating and buying so much stuff—prevention is worth pounds of consumption.

By the way, it was not only soft drinks and beer that used returnable containers. Postwar development of milk cartons and one-way plastic jugs, contributed to the decline of refilling bottles in the milk industry. The one-way milk containers began with the paper carton in 1906 and progressed to the plastic-coated paper version in 1932, and in 1964, the milk industry ushered in one-way plastic jugs made of high-density polyethylene (HDPE). In the 1970s and 1980s, refillable bottles made of polycarbonate plastic were used in schools and in other institutions.

WHY WE LITTER

For millennia, it really didn't matter what people tossed on or under the ground because it was all biodegradable. The real problem came along with synthetic material, especially plastics, and throwaway paper products that people began to toss away because it was easy.

Most people litter when they think they're not being watched—like picking your nose while in a car, or hastily tossing a cigarette butt or gum wrapper on the sidewalk. When people think they are being watched a different set of socialized conventions take over. And it's not a lot more sophisticated than monkey see, monkey do.

"One of the things that's fundamental to human nature is that we imitate the actions of those around us," wrote Robert Cialdini, emeritus professor of psychology and marketing at Arizona State University and author of *Influence: The Psychology of Persuasion*. He has conducted a number of landmark studies in littering and litter prevention—all of them pointing to the fact that people are likely to do what they think is expected of them. Change these, claims Cialdini, and you'll change people's behavior. When in an environment that is littered, people will toss trash; if there is none, people are significantly less likely to litter.

This research echoes the "broken windows/vandalized car theory," by social scientists James Q. Wilson and George L. Kelling. This theory holds that people are more likely to break windows, write graffiti, deface an environment, and trash or strip cars if it's already been started. It's like a feeding-frenzy response.

People also litter because they do not feel responsible for public areas like streets and parks, especially when not in "their" immediate neighborhood. And the more they litter, the more it becomes a habit, and the worse the community looks. People usually litter outside their own environs where their trash becomes someone else's problem, believing somebody else—a maintenance worker or responsible neighbor—will pick up after them.

One of the cures for littering is education and awareness. Remember when recycling garbage was such a hassle? But if you don't do it, these days you're considered a public enemy suitable for shaming.

A GREAT EXAMPLE OF GOOD INTENTIONS

Eco-warrior Nadia Sparkes, a 12-year-old from Norwich, Connecticut, decided to literally take matters into her own hands. The high schooler has been picking up trash along the two-mile route from her school to her home for months now, hauling it in the basket of her bike. Despite her green intentions, some of the kids at Nadia's school have dubbed her "Trash Girl" and have bullied her for her righteous efforts.

"I'm doing something to protect the world they also live in. It's everyone's job. We are all responsible for keeping this world safe, instead of believing that it's always someone else's job." She has currently been elevated to cartoon superhero by a media outlet—and a movie might even be in the works.

Cathy Ives publishes Green Eco Services, a great website about helping to care for our Earth. Hers is a quest that started in her backyard and focuses on the mess on the beaches in San Diego, and she tries to make a buck or two on the trash she picks up in the process. It's a double win.

"The Mission Bay Park is over 4,235 acres, including both water and land," Ives says. "I am only picking up less than a mile of the 27 miles of shoreline. My beach cleaning focus is what we call the 'Tide To Towel' area. I do try to pick up at least 100 pieces of plastic in the parking lot (close to the beach) and 100 pieces of plastic in the picnic area by the Belmont Park parking lot." She has picked up more than 1,700 pounds of trash over the last several years. The goal is to make a living—by recycling, reusing, and selling off what people throw away, and by doing more than her part for the beaches in her home.

SOURCES

"A Garbage Timeline." Rotten Truth about Garbage, http://astc.org/exhibitions/rotten/timeline.htm.

Derbyshire, David. "How Litter and Graffiti Can Poison Our Minds by Turning Us More Racist and Homophobic." *MailOnline*, April 8, 2011, http://www.dailymail.co.uk/sciencetech/article-1374673/Litter-graffiti-make-people-racist-homophobic.html#ixzz5D3AgHQY3.

Environmental Defense Fund. "*Methane Research: The 16 Study Series: An Unprecedented Look at Methane from the Natural Gas System*. New York: Environmental Defense Fund, April 2017, https://www.edf.org/sites/default/files/methane_studies_fact_sheet.pdf.

Environmental Protection Agency. "Resource Conservation and Recovery Act (RCRA) Laws and Regulations." https://www.epa.gov/laws-regulations/summary-resource-conservation-and-recovery-act.

Franklin, Pat, et al. "Plastic Water Bottles Should No Longer Be a Wasted Resource." Container Recycling Institute Website (2010), http://www.container-recycling.org/index.php/issues/bottled-water.

Gatens, Hannah. "Waste Not Want Not." *W&M* (2015), http://wmalumnimagazine.com/2015/summer/tribe/waste-not-want-not/.

Kiss, Lisa. "The Truth about Trash—The Throw-Away Society." *Planet Thoughts*, March 11, 2008, http://www.planetthoughts.org/?pg=pt/Whole&qid=1944.

"Land of Waste: American Landfills and Waste Production." Save on Energy, https://www.saveonenergy.com/land-of-waste/.

Royte, Elizabeth. *Garbage Land: On the Secret Trail of Trash*. New York: Back Bay Books, 2006.

Sasko, Claire. "Only One State Has More Trash Per Capita Than Pennsylvania." *Philadelphia*, August 4, 2016. https://www.phillymag.com/news/2016/08/04/pennsylvania-landfills/.

Sheridan, Dick. "Trash Fight." *New York Daily News*, http://www.nydailynews.com/new-york/trash-fight-long-voyage-new-york-unwanted-garbage-barge-article-1.812895.

Short, Aaron. "New York Is on Top of the Heap in Garbage Hauling Costs." *New York Post*, May 24 2017, https://nypost.com/2014/05/24/new-york-is-top-of-the-heap-in-garbage-hauling-costs/.

"Top Ten Countries That Produce the Most Trash." ILLYear4, https://sites.google.com/site/iilyear4/top-10-countries-that-produce-the-most-waste.

Tulipano, Rachael. "America the Wasteful: A Detailed Look into Our Throw-Away Society." *GenTwenty*, https://gentwenty.com/america-the-wasteful-a-detailed-look-into-our-throw-away-society/.

"Use It and Lose It: The Outsize Effect of U.S. Consumption on the Environment." *Scientific American*, https://www.scientificamerican.com/article/american-consumption-habits.

Wagner, Vivian. "Littering and Following the Crowd." *Atlantic Monthly*, August 1, 2014, https://www.theatlantic.com/health/archive/2014/08/littering-and-following-the-crowd/374913/.

"Wetlands." *Home Reference Encyclopedias*, Fall 2005, almanacs, transcripts, and maps, https://www.encyclopedia.com/reference/encyclopedias-almanacs-transcripts-and-maps/wetlands.

Wrona, Nicole. "Life after the Landfill." *Waste Dive*, March 28, 2014. https://www.wastedive.com/news/life-after-the-landfill-sites-find-new-life-when-repurposed/244568/.

Zhang, Sarah. "Charts: What Your Trash Reveals about the World Economy." *Mother Jones*, July 2012, https://www.motherjones.com/environment/2012/07/trash-charts-world-bank-report-economy/.

2

WASTING OUR O_2

Our lives begin and end with a breath of air.

The way we trash our air, you'd think it was the most abundant element on Earth—and you'd be right. You might also think that because it is the most plentiful that it's OK to waste it. You'd be dead wrong. This invisible element is the only thing that you can live only moments without. And we're ruining it at an atmospheric rate.

Earth is the only planet in the solar system with an atmosphere that can sustain abundant life, and the most abundant element in the Earth's crust is oxygen (O_2), making up 46.6 percent of the Earth's mass, and 21 percent of our atmosphere. It is part of the blanket of gases that not only contains the air that we breathe but also forms a shielding layer of ozone (O_3) in the atmosphere that screens out damaging ultraviolet radiation from the Sun, warms the planet by day, and cools it at night.

Somewhere around 2.7 billion years ago our primordial benefactors, tiny organisms known as cyanobacteria, blue-green algae, produced oxygen as a waste product of photosynthesis. It took billions of years for oxygen to accumulate and beget the ascendance of the only animal species that creates compounds, gases, and waste that nature can't biodegrade.

Even if humans quit polluting the atmosphere by reducing carbon dioxide release right now, it still could take more than a century before the air clears.

ECO-AIR CONTAMINATION

Although Mother Nature causes natural pollution via wildfires, volcanic eruptions, wind erosion, pollen dispersal, decomposing of organic compounds, biological animal body functions, and natural radioactivity, *Smithsonian* magazine reports that more than 80 percent of air pollution is caused by man.

Volcanic eruptions cause enormous amounts of matter, toxic and otherwise, to be spewed on land and hurled into the atmosphere. These materials can lead to deadly acid rain, as well as ash that disrupts car and air travel and a lot of other activities—like breathing.

Dust and dirt particles whipped up by winds are sometimes strong enough to become dangerous storms, and although not toxic they are capable of causing a slew of respiratory diseases in human beings.

Then there is something called black carbon. For centuries, humans have engaged in activities that produce black carbon particles that are released into the atmosphere in the form of smoke—cooking with biomass, combusting petro-fuels, burning trees, and emitting gaseous exhaust. When black carbon particles reach the atmosphere, they form a heat-absorbing layer that causes temperatures to rise. Raindrops tend to form around black carbon particles in the atmosphere, and when they fall to the ground, they absorb heat there, too, thus magnifying their warming effect, and end up in glaciers, causing them to melt faster.

Farm animals during digestive cycles release toxic methane into the atmosphere that is flammable, can damage the ozone layer, and is a very potent greenhouse gas that traps heat in the atmosphere, exacerbating climate change.

Nuclear elements like uranium decompose and release radon into the atmosphere. The gas is highly radioactive and it can cause serious health damage to people who breathe it. After smoking, radon intake is the second-largest contributing factor to lung cancer. Because it is invisible and odorless measures to detect the release of radon, like gauges in the home, are a good idea.

Anyone who suffers from allergies or asthma can testify to the result of pollen released by plants and trees. The spread of pollen through the air helps to fertilize plants but brings misery to seasonal allergy and asthma sufferers—altogether more than 75 million Americans.

ANTHROPOGENIC AIR POLLUTION

Corrupting the atmosphere can be defined as injecting toxic chemicals or compounds into the air at levels that pose a health risk to living organisms and the planet in general. Anthropogenic pollution is that caused by human activities such as the combustion of fossil fuels, mining, construction, transportation, industrial work, smelting, agriculture, waste incinerators, as well as home furnaces.

Anthropogenic air pollution is typically separated into two categories: indoor and outdoor. Indoor air pollution involves exposure that takes place inside of the "built" environment (structures that house us), and inside those buildings with a combo of natural and man-made contaminants. The average American spends something like 90 percent of his or her time indoors. This means that we had better pay attention to indoor air quality because it can be 10 to 100 times worse than the air outside.

Primary sources of indoor pollution are carbon monoxide, an odorless, colorless gas produced by burning petroleum products; secondhand tobacco smoke, composed of more than 3,800 different chemical compounds; and radon, a radioactive gas that can seep into homes through cracks in the foundation, floors, and walls. "Off-gassing" (a release of chemicals from various substances, like the new car or carpet smell) can come from refrigerators, air conditioners, humidifiers, household cleaners, degreasers, paints, and glues, as well as from newly installed carpets, flooring, particle board, paneling, mattresses, and furniture. Then there are pesticides from indoor or outdoor plants. Biological sources can come from plants, animals, and insects; gas, decomposing skin and hair cells; and various molds.

The health effects of exposure to volatile organic compounds (VOCs) vary depending on the person, the chemical, and its concentration. Sometimes, off-gassing causes temporary dizzyness after the use of cleaning products or paint, but most of the time you feel nothing. In the long term, however, doctors are concerned about continuing exposure to off-gassing, as chemical contaminants have been linked to 180 diseases. Because there can be dozens of items in your home at various stages of off-gassing, it's a good idea to know what they are (ask when you buy and look at labels), and use an air purifier to filter the VOCs out of the air when using them in bulk.

Air pollution is not new, as it began when we first started using fuel and fire for domestic use. But the unbridled expansion of industry over the past two centuries has resulted in changes in the chemical composition of the atmosphere on a scale unprecedented in human history—twice the amount of CO_2 has been created in the last 40 years as during the previous 150 years.

EPA Protection of Air

The Clean Air Act of 1970, which set limits on emissions and standards for air quality, provided government funding for pollution control research, and made it possible for citizens to sue those who violated the standards. The standards were set forth by the Environmental Protection Agency (EPA), which identified six pollutants as those posing the greatest threat to human health:

- particle pollution (particulate matter)
- ground-level ozone
- carbon monoxide
- sulfur oxides
- nitrogen oxides
- lead

Pollution Points

Point source pollution is the contaminants that come from a single source, such as smokestacks at a single factory. Nonpoint source pollution is what comes from many sources, such as all of the transportation vehicles in the United States.

Primary pollutants are those that cause direct harm in the atmosphere. Secondary pollutants are those harmful substances that are created from the reactions between primary pollutants and the components of the atmosphere, like water vapor and CO_2 contributing to global warming.

These foul taints can be severe or deadly health risks even in the small amounts that we breathe. Almost 200 are regulated by law—some of the most common are mercury, lead, dioxins, and benzene. These are most often emitted during natural gas, gasoline, or coal combustion.

The by-products of burning coal releases nitrogen oxides, sulfur dioxide, particulate matter (PM), mercury, and dozens of other substances known to be hazardous to human health. Benzene, released by gasoline, is classified as a carcinogen by the EPA, and it can cause eye, skin, and lung irritation in the short term as well as blood disorders.

Dioxins, a highly toxic compound produced as a by-product in some manufacturing processes, present in small amounts in the air; can affect the liver; and harm the immune, nervous, and endocrine systems, as well as reproductive functions. Lead in large amounts can damage children's brains and kidneys, and even in small amounts it can affect children's IQ and ability to learn. Mercury, called the "Mad Hatter" chemical, affects the central nervous system.

In one recent study, research suggests that a pregnant mother's exposure to air pollution may increase her baby's odds of developing slower brain processing speeds and onset of attention deficit disorder (ADD).

Petro Pollution

Smog, or "ground-level ozone," occurs when emissions from anything that combusts fossil fuels reacts with sunlight and soot made up of tiny particles of chemicals, soil, smoke, dust, or allergens, in the form of gas or solids, that are carried in the air. In many parts of the country pollution has reduced the distance and clarity of what we see by 70 percent.

Smog can irritate the eyes and throat and also damage the lungs—especially of people who work or exercise outside, children, and senior citizens, and it's worse for people who have asthma or allergies.

CO_2 and HFCs

According to a 2014 EPA study, carbon dioxide was responsible for 81 percent and methane 11 percent of greenhouse gases in the atmosphere. Another class of greenhouse gases, hydrofluorocarbons (HFCs), manufactured for use in refrigeration, air conditioning, foam, aerosols, fire protection, and solvents, are thousands of times more powerful than carbon dioxide in their ability to trap heat in the air. The National Resources Defense Council (NRDC) estimates that the agreed HFC phase-down will avoid the equivalent of more than 80 billion tons of CO_2 over the next 35 years.

A Venomous Variety of Vapors

Carbon monoxide is odorless and colorless (except when mixed with particulates or burning petro-fuels), but it is also produced from forest fires and the burning of industrial waste. It's dangerous to humans because even in very small concentrations, it can prevent oxygen from being delivered to major organs in the body. Breathe enough, and it can kill.

Sulfur dioxide is also a colorless gas; 70 percent comes from combustion at coal-fired industrial power plants. Once in the air, it may react further to create sulfuric acid, which can fall back to Earth as acid rain. Sulfur dioxide can also cause respiratory illnesses when it's breathed.

Nitrogen oxide has a sharp, sweet smell. Both nitric oxide and nitrogen dioxide are in this family and contribute to wet and dry acid rain. These gases contribute to the smoggy brown haze you see over large cities like Los Angeles. More than half of these pollutants come from vehicle combustion emissions and industrial combustion, and they can cause respiratory illnesses.

The poisonous precipitation called acid rain can cause a variety of environmental effects. Both the wet and dry variety can be carried by the wind, sometimes for very long distances. It falls on buildings, cars, and vegetation, killing plants, and corrupting waterways. In dry form, it can be inhaled and can cause respiratory as well as other health problems.

POPs and PMs

Persistent organic pollutants (POPs) and PMs are organic compounds that are resistant to environmental degradation and used as pesticides, solvents, pharmaceuticals, and industrial chemicals. They can have potentially significant impacts on human health and the environment. PMs are tiny particles of solid or liquid suspended in the air from fires, dust, fly ash, soot, smoke, aerosols, fumes, mists, and condensed vapors that are suspended in Earth's atmosphere. POPs are a group of chemicals that are persistent in the environment and travel vast distances via air and water, are very toxic, and can cause cancer and other adverse health effects in animals and humans.

Air Pollution Per Person

An average family in the United States causes the following amounts of pollution in the air each year:

- CO_2 = 85 tons
- Ozone causing pollution (NOx) = 325 pounds
- Acid rain (SO_2) = 411 pounds
- Small particulates = 43 pounds
- Toxic lead pollution (Pb) = 1.2 ounces
- Toxic mercury pollution (Hg) = 0.04 ounces

To find out how much pollution is caused by using household electricity check out the Cleaner and Greener Pollution Calculator (http://www.cleanerandgreener.org/resources/pollutioncalculator.html).

HEALTH CONCERNS

Air pollution costs more than $200 billion a year and contributes to 16,000 premature births in the United States each year, reports the *Journal of Environmental Health Perspectives* by Dr. Leonardo Trasande, an associate professor at New York University's Langone Medical Center. "The bulk cost of emissions is the result of its health impacts," says Paulina Jaramillo, an assistant professor of engineering and public policy at Carnegie Mellon University. However, Jaramillo also states that with the declining use of coal for energy, the costs of air pollution is declining.

Nevertheless, there is still a lot of pollution to contend with. One place to look for evidence is in the health of lichens. Sentinel species are like organic body guards—plants and animals that are used to judge changes in the environment. For example, canaries were used by early miners to indicate their quality of air—if the birds died it was time to leave. Lichens, a slow-growing plant that can be found on rocks, walls, and trees are like those canaries. Undisturbed lichens can live for hundreds of years, but in polluted air it dies. Parts of England have already lost 89 percent of their lichens. Southern California has lost 50 percent.

Besides lichens, air pollution kills more than 2 million people each year worldwide, according to a study published in the journal of *Environmental Research Letters*, and is responsible for 34 percent of strokes, 27 percent of heart disease, and 36 percent of lung disease worldwide, according to the World Health Organization (WHO).

Many studies link air pollution to a broad range of ailments. Researchers are now finding that more than the lungs are at risk because dirty air may be an accomplice to diabetes, obesity, coronary disease, optical dysfunctions, and even forms of organic brain disorders and dementia.

The effects of air pollution on human health can vary widely. The very young, the old, and those with vulnerable immune systems are most at risk from air pollution. If the pollutant is highly toxic, the effects on health can be widespread and severe. An irritant less than 2.5 micrometers (about a quarter of the width of the smallest grain of pollen) may cause respiratory illnesses and cardiovascular disease as well as an increase in the seriousness of asthma attacks.

Scientists now know that inhaling pollutants triggers a flurry of physiological coping mechanisms throughout the body. "Until 20 years ago, we thought that air pollution affected only the respiratory system," says Petros Koutrakis, an environmental chemist who heads the EPA Harvard Center for Ambient Particle Health Effects. The American Heart Association published a statement in *Circulation* laying out "a strong case that air pollution increases the risk of cardiovascular disease," the leading cause of deaths in America.

Indoor air pollution can increase a person's chances of having flares of chronic lung problems, such as asthma or chronic obstructive pulmonary disease (COPD). There are also likely longer-term effects from ongoing exposure that are more difficult to measure, such as the likelihood of lung cancer from radon exposure, as well as second hand and third hand smoke, and other chronic respiratory problems of the nose, throat, eyes, and lungs.

Eyes on Air Pollution

The most frequently reported vision problem linked to air pollution is dry eyes. This is especially common where there is smog or smoke, or when dust or particles are in the air, basically almost everywhere.

There is a spectrum of optical symptoms occurring due to air pollution. It may range from simple irritation and burning to severe allergy, cataracts, and even cancer. The most common problems are:

- redness.
- burning, itchy sensation.
- watering, ropey, sticky discharge.
- dry, gritty feeling.
- blurred vision.
- photophobia (light sensitivity).

If eyes become irritated, don't rub them—but do this:

- Place a cool compress to closed eyes.
- Use lubricating eye drops provided by eye specialists.
- Wear sunglasses outdoors.
- Do not splash water into open eyes.
- Avoid contact lens and eye makeup when eyes are sore.
- Rest the eyes.

Asthma

Here are some facts on asthma:

- 40,000 people to miss school or work.
- 30,000 people have an asthma attack every day, and 11 people die.
- Senior citizens in the United States account for nearly 2,400 (60 percent) of more than 4,000 deaths due to asthma annually.
- More females die of asthma than males, and women account for nearly 65 percent of asthma deaths overall.
- Asthma accounts for two million emergency room visits and accounts for more than 10 million outpatient visits and 500,000 hospitalizations in the United States each year.

- 44 percent of all asthma hospitalizations are for children, the third-ranking cause of hospitalization and the number one cause of school absenteeism.
- Since 1980, asthma death rates have increased more than 50 percent among all genders, age groups and ethnic groups, and the death rate for children under 19 years old has increased by nearly 80 percent.
- The annual cost of asthma is estimated to be nearly $18 billion.

Brain Drain

In a June 2014 study conducted by researchers at the University of Rochester Medical Center and published in the journal *Environmental Health Perspectives*, researchers found early exposure to air pollution causes the same damaging changes in the brain as autism and schizophrenia. The study also shows that air pollution also affected short-term memory, learning ability, and impulsiveness.

In 2015, researchers at Harvard University and Syracuse University reported test subjects having trouble remembering, learning new things, concentrating, or making decisions that affect everyday life, caused by impurities in indoor air. The tests measured brain-based skills needed to carry out any task from the simplest to the most complex.

Participants' performance in controlled laboratory atmospheres that simulated those found from "conventional" to "green" buildings, was evaluated. Lower scores were observed where there were increased concentrations of either VOCs or CO_2. The highest impurity levels tested were in some classroom and office environments.

In 2016 in *Environmental Health Perspectives*, researchers reported that people with Alzheimer's and other forms of dementia also experienced heavy exposure to air pollution. Danish researchers, with colleagues in the United States and Taiwan, also published a study in *Environmental Health Perspectives* studying people with and without Parkinson's disease and their exposure to nitrogen dioxide, a marker for traffic-polluted air. The scientists identified 1,828 people in Denmark with Parkinson's and compared them with approximately the same number of randomly selected healthy people. It revealed that those exposed to the highest levels of air pollution also had a great risk of developing the disease.

The number of people who die in America every year due to air pollution is more than 50,000, and 80 percent of lung diseases are caused due to pollution from transportation vehicles.

One study, reported in the *Journal of the American Medical Association* (JAMA), followed 4,602 children in Southern California between 1993 and 2012 to see how lung health correlated with air pollutants. As levels of ozone, nitrogen dioxide, and particulate matter fell over time, so did the number of children who reported a daily cough, persistent congestion, and other symptoms of irritated lungs. In communities with the greatest drop in pollutants during the study period, bronchitis prevalence fell by as much as 30 percent in children with asthma.

Even with vast improvements in air quality since the 1970s, people haven't stopped dying from the air they breathe. An analysis published in 2013 from researchers at the Massachusetts Institute of Technology (MIT) estimated that about 200,000 premature deaths occur each year in the United States because of fine particulates in air pollution. A study published in *Environmental Health Perspectives* reported that daily deaths over a decade in Boston peaked on days when concentrations of air pollutants were at their highest, even though those levels would currently satisfy the EPA as being safe.

According to the U.S. National Library of Medicine, National Institutes of Health research has demonstrated an increased risk of developing asthma and COPD, which makes it hard to breathe, from increased exposure to traffic-related air pollution. Additionally, air pollution has been associated with increased hospitalization and mortality from asthma and COPD including chronic bronchitis and emphysema.

Cancer

In the American Cancer Society's (ACS) study, a half million men for at least 10 years were classified into various categories by place of residence and whether or not they were occupationally exposed to dusts, fumes, vapors, and habits including smoking. Men who said they were occupationally exposed had mortality rates of lung cancer 14 percent greater than the nonexposed. Among those not exposed, there were little or no differences in mortality ratios by urban or rural place of residence or whether they lived in cities with high, medium, or low levels of suspended particulate matter or benzene organic matter. Living close to busy traffic appears to be associated with elevated risks in lung cancer deaths and cardiovascular deaths.

Pounds and Pollution

Lately, studies of air pollution have moved into unexpected territory and turned up compelling evidence that air quality may contribute to being overweight. A 2014 study by Frank Gilliland, an environmental epidemiologist at the University of Southern California in Los Angeles and colleagues studied body mass index (BMI), a value derived from the weight and height of an individual, among children exposed to traffic-related air pollution. Children from the most exposed areas had a 14 percent larger BMI increase.

Adults also appear to be affected. Researchers from Harvard Medical School published a study in 2016 in the journal *Obesity* observing whether adults living with constant exposure to traffic are more likely to be overweight. People who lived within 60 meters of a major road had a higher BMI, and more fat tissue than those who lived 440 meters from a busy road.

Links to Diabetes

Gilliland and colleagues published data in *Diabetes* magazine finding links between air pollution and diabetes in children. In the study, 314 overweight or obese children in Los Angeles were followed for an average of three years. At the end of the study, children who lived in neighborhoods with the highest concentrations of nitrogen dioxide and particulates (air traffic pollution) had signs of impaired pancreatic beta cells, which produce insulin. This can also produce a rapid decline in memory and other mental skills in old age even among people who don't have diabetes.

Simple Steps

The following are some basic steps to take to protect yourself and while you're at it the Earth as well.

- When pollution levels are high, limit the time for a jog and keep windows closed. Generally, ozone levels tend to be lower in the morning.
- When exercising outside, stay away from heavily trafficked roads. Then shower and wash your clothes to remove fine particles.
- Use sunscreen. When ultraviolet radiation comes through the weakened ozone layer, it can cause skin damage and skin cancer.
- Avoid smoking.
- Use craft supplies in well-ventilated areas.
- Make sure gas appliances are well-ventilated.
- Have car emissions tested regularly.
- Reduce the number of trips you take in your car. Electric vehicles produce less air pollutants.
- Minimize air fresheners (they can have unhealthy ingredients, same for scented candles).
- Test your home for radon, and carbon monoxide with detectors.
- Dust surfaces and vacuum frequently.
- Make sure exhaust fans are functioning in your bathrooms and kitchen.
- Choose environmentally friendly cleaners.
- Use environmentally safe paints and cleaning products whenever possible, and seal containers of chemicals.
- Advocate for emission reductions from power plants and more stringent national vehicle emission standards.
- Reduce or eliminate fireplace and wood stove use and avoid burning leaves, trash, and other materials.
- Consider using gas logs instead of wood.
- Avoid using gas-powered lawn and garden equipment.

POLITICAL FIGHT AGAINST FOSSIL FUELS

The EPA announced in 2017 that the department had signed a measure to repeal President Barack Obama's policy to curb greenhouse gas emissions from power plants. The Obama "Clean Power Plan" would have pushed states to move away from coal in favor of sources of electricity that produce fewer carbon emissions. The repeal proposal supposedly fulfills a promise President Trump made to eradicate his predecessor's environmental legacy, denouncing it as a job-killing regulation. But air pollution kills people. Some states with big pollution problems, notably California, have taken it upon themselves to follow stricter environmental protection policies

The top 10 areas of air pollution in the United States are

- Bakersfield, California
- Visalia–Porterville–Hanford, California
- Fresno–Madera, California
- Modesto–Merced, California
- Fairbanks, Alaska
- San Jose–San Francisco–Oakland, California
- Salt Lake City–Provo–Orem, Utah
- Logan, Utah
- Los Angeles–Long Beach, California
- Reno–Carson City–Fernley, Nevada

While the repeal of the Clean Power Plan offers a reprieve for America's coal industry, it is unlikely to halt the decline of coal altogether, as the use of coal is declining and many power plants have already shifted to natural gas and alternative forms of energy.

Trump also took the United States out of the Paris Accord, which promised to reduce overall pollution and included measures to help developing countries refrain from deforestation, among other projects, that would have decreased the amount of CO_2 in the atmosphere.

SOURCES

"17 Simple Ways to Prevent Air Pollution in Your Home." *Health Essentials*, https://health.clevelandclinic.org/17-simple-ways-prevent-air-pollution-home/.

"Abundance of the Chemical Elements." Wikipedia, https://en.wikipedia.org/wiki/Abundance_of_the_chemical_elements#Earth.

"Air Pollution Causes 200,000 Early Deaths Each Year in the U.S." MIT Laboratory for Aviation and Environment, August 29, 2013, http://lae.mit.edu/2013/08/29/air-pollution-causes-200000-early-deaths-each-year-in-the-u-s/.

"Air Pollution Facts." Conserve Energy Future, https://www.conserve-energy-future.com/various-air-pollution-facts.php.

"Air Pollution from Coal-Fired Power Plants." SourceWatch, the Center for Media and Democracy, https://www.sourcewatch.org/index.php/Air_pollution_from_coal-fired_power_plants.

"Air Pollution from Particulate Matter." Texas Commission on Air Quality, https://www.tceq.texas.gov/airquality/sip/criteria-pollutants/sip-pm.

"Air Pollution, How Do We Cause Air Pollution." Leaner and Cleaner, http://www.cleanerandgreener.org/resources/air-pollution.html.

Air Pollution Killing over Two Million Annually." CNN, July 16, 2013, https://www.cnn.com/2013/07/16/world/air-pollution-killing-study/index.html.

"Asthma." National Heart, Blood, Lung Institute, https://www.nhlbi.nih.gov/health-topics/asthma.

"Asthma Facts and Figures." Allergy and Asthma Foundation of America, http://www.aafa.org/page/asthma-facts.aspx.

Biel, Laura. "The List of Diseases Linked to Air Pollution Is Growing." *Science News*, September 2017, https://www.sciencenews.org/article/list-diseases-linked-air-pollution-growing.

Biello, David. "The Origin of Oxygen in Earth's Atmosphere." *Scientific American*, August 19, 2014, https://www.scientificamerican.com/article/origin-of-oxygen-in-atmosphere/.

"Black Carbon Pollution Emerges as Major Player in Global Warming." *Science Daily*, March 24, 2008, https://www.sciencedaily.com/releases/2008/03/080323210225.htm.

"Causes of Air Pollution." *SaveEarth*, https://www.saveearth.info/air-pollution/.

"COPD and Air Pollution." COPD.net, https://copd.net/basics/causes-risk-factors/who-is-at-risk/air-pollution/.

Doniger, David. "Countries Adopt Kigali Amendment to Phase Down HFCs." Natural Resources Defense Council, October 14, 2016, https://www.nrdc.org/experts/david-doniger/countries-adopt-kigali-amendment-phase-down-hfcs.

"Environmental Science Second Year." *Quizlet*, https://quizlet.com/204294314/environmental-science-second-year-flash-cards/.

Freedman, Andrew. "The Last Time CO_2 Was This High, Humans Didn't Exist." *Climate Central*, May 2, 2013, http://www.climatecentral.org/news/the-last-time-co2-was-this-high-humans-didnt-exist-15938.

Friedman, Lisa, and Brad Plumer. "E.P.A. Announces Repeal of Major Obama-Era Carbon Emissions Rule." *New York Times*, October 9, 2017, https://www.nytimes.com/2017/10/09/climate/clean-power-plan.htm.

Gathania, Rishav. "How Does Air Pollution Affect Global Warming?" *Quora*, May 11, 2017, https://www.quora.com/How-does-air-pollution-affect-global-warming.

Gilliland, Frank D., et al. "A Longitudinal Cohort Study of Body Mass Index and Childhood Exposure to Secondhand Tobacco Smoke and Air Pollution." *Environmental Health Perspectives*, May 2013, https://ehp.niehs.nih.gov/1307031/.

Green, Jared. "Six Ways Human Activity Is Changing the Planet." The Dirt, April 6, 2010, https://dirt.asla.org/2010/04/06/six-ways-that-human-activity-is-changing-the-planet/.

Hammond, E. Cuyler. "General Air Pollution and Cancer in the United States." *Preventive Medicine* 9, no. 2 (1980): 206–11.

"Help Save Nature." https://helpsavenature.com/natural-causes-of-air-pollution.

Kazmeyer, Milton. "Human Impact on the Earth's Atmosphere." *Sciencing*, January 30, 2018, https://sciencing.com/human-impact-earths-atmosphere-3677.html.

Khader, Amin Burhari Abdul. "Air Pollution." Omtex Classes, Scribd, https://www.scribd.com/document/9520178/Air-Pollution.

"Learning Space, Ventilation." WikiVisually, https://wikivisually.com/wiki/Learning_space#Ventilation.

Mackenzie, Jillian. "Air Pollution: Everything You Need to Know." Natural Resources Defense Council, November 1, 2016, https://www.nrdc.org/stories/air-pollution-everything-you-need-know.

"Outgassing." Wikipedia, https://en.wikipedia.org/wiki/Outgassing.

Oudin, Anna, et al. "Traffic-Related Air Pollution and Dementia Incidence in Northern Sweden: A Longitudinal Study." *Environmental Health Perspectives* 124, no. 3 (2016): 306–12.

"Particle Pollution." Air Now, https://airnow.gov/index.cfm?action=aqibasics.particle.

"Pollen Allergy." New York Allergy & Asthma Experts, http://allergyexperts.com/pollen-allergy/.

"Polycyclic Aromatic Hydrocarbons (PAHs), Factsheet." *Centers for Disease Control and Prevention*, April 7, 2017,https://www.cdc.gov/biomonitoring/PAHs_FactSheet.html.

Rao, Muhammad Moon. "An Introduction to Environmental Chemistry." Slideshare, www.slideshare.net/MuhammadFazalurRehma3/.

Rapaport, Lisa. "Premature Births Linked to Air Pollution Cost More Than $4 Billion a Year." *Huffington Post*, March 29, 2016, https://www.huffingtonpost.com/entry/premature-births-linked-to-air-pollution-cost-more-than-4-billion-year_us_56face40e4b0143a9b4974e6.

Riddell, Jennifer. "Tracking Lichen Community Composition Changes Due to Declining Air Quality over the Last Century in Southern California." *Bibliotecha Lichenologica* 106 (2011): 263–77, https://www.fs.fed.us/pnw/pubs/journals/pnw_2011_riddell001.pdf.

Rodden, Janice. "Air Pollution Linked to ADHD." *ADDitude*, November 13, 2014, https://www.additudemag.com/air-pollution-linked-to-adhd/.

Savage, Sam. "Americans Spend 90 Percent of Time Indoors—Exposed to Indoor Allergens." *Red Orbit* (2005), http://www.redorbit.com/news/health/244736/americans_spend_90_ percent_of_time_indoors__exposed_to.

"The Most Polluted Cities in America." *USA Today*, https://www.usatoday.com/story/money/business/2015/08/22/24-7-wall-st-most-polluted-cities/32130565/.

Shear, Michael. "Trump Will Withdraw U.S. from Paris Climate Agreement." *New York Times*, June 1, 2017, https://www.nytimes.com/2017/06/01/climate/trump-paris-climate-agreement.html.

Solomon, S., J. S. Daniels, T. J. Sanford, G. K. Plattner, R. Knutti, and P. Friedlingstein. "Persistence of Climate Changes Due to a Range of Greenhouse Gases." *Proceedings of the National Academy of Sciences* 107, no. 43 (2010).

"Sources of Indoor Air Pollution." *Carbon Monoxide Kills* ,http://www.carbonmonoxidekills.com/are-you-at-risk/sources-of-indoor-air-pollution/.

"Strong Link between Diabetes and Air Pollution Found in National U.S. Study." *Science Daily*, September 30, 2010, https://www.sciencedaily.com/releases/2010/09/100929105654.htm.

"The Facts about Air Quality and Coal-Fired Power Plants." Institute for Energy Research, http://instituteforenergyresearch.org/studies/the-facts-about-air-quality-and-coal-fired-power-plants/.

U.S. Environmental Protection Agency. "Overview of Greenhouse Gases." https://www.epa.gov/ghgemissions/overview-greenhouse-gases.

Vuong, Zen. "Air Pollution Linked to Heightened Risk of Type 2 Diabetes." *Medical XPress*, February 7, 2017,https://medicalxpress.com/news/2017-02-air-pollution-linked-heightened-diabetes.html.

"What Air Pollution Does to Eyes and Vision." Rebuild Your Vision, https://www.rebuildyourvision.com/blog/interesting-vision-facts/what-air-pollution-does-to-eyes-and-vision/.

"What Are the Six Common Air Pollutants." *Eco Link News*, https://ecolink.com/info/six-common-air-pollutants/.

"What Is Air Pollution? Definition, Sources & Types." *Study.com*, https://study.com/academy/lesson/what-is-air-pollution-definition-sources-types.html.

3

WATERWAYS WASTE WATCH

We forget that the water cycle and the life cycle are one. The water on which all life depends has become a global garbage can.
—*Jacques Yves Cousteau*

THE OCEANS AT RISK

It is the place where life on Earth was born. It covers 71 percent of the planet, produces more than half of the oxygen in the atmosphere, and absorbs about 40 percent of the 2.4 million pounds of CO_2 generated every second. Around 80 to 90 percent of plastic pollution enters from our land and rivers into the hydrosphere dump that has become the largest marine trash vortex on the planet.

In the past, we thought that because of their immensity we could never take too much out of our seas or put too much in. It was a mistake of truly immense proportions. Imagine an area 34 times the size of Manhattan (302.6 square miles) covered ankle-deep in plastic waste as far as the eye can see. That's about how much rubbish and debris ends up in our oceans every year, according to Jenna Jambeck, an environmental engineer, who studies these things.

Of the 260 million tons of plastic the world produces each year, about 8 to 10 percent ends up in the oceans, according to a Greenpeace report. Seventy percent of that eventually sinks, damaging life on the ocean floor, threatening at least 600 different species, and the creation of oxygen, reports the Ocean Conservancy.

The largest concentrations of garbage on the planet are in the oceans. Called gyres—rotating currents of water and trash—there are five major ones: the North Atlantic, South Atlantic, South Pacific, Indian Ocean, and the largest is the North Pacific. Its size changes with currents and winds, stretching roughly from the west coast of North America to the east coast of Japan. Contrary to popular buzz, it's not an enormous island of solid floating trash, but rather a boundless, almost immeasurable soup of plastic rubble. It is impossible to quantify the total amount of plastic in the ocean, but the latest figures estimate there are up to 51 trillion particles, or "500 times more than the stars in our galaxy" according to the United Nations Environment Program. The U.S. National Oceanic and Atmospheric Administration (NOAA) has said that marine debris "has become one of the most pervasive pollution problems facing the world's oceans and waterways."

Since its discovery in 1997 the Great Pacific Gyre has become the stuff of fact and fable, and current estimates of its size has it spanning from 160,000 to 386,000 million square miles. An aerial survey last year found it is far bigger than previously estimated, and it is now clearly visible from space.

Researchers have found 750,000 microplastic pieces per square kilometer (one mile is 1.6 kilometers) in the Pacific Gyre, and the Ellen MacArthur Foundation (started by the competitive round the world sailor) estimates that there will be more plastic than fish in the oceans by 2050.

Food Faux Pas = Plastic Death

In July 2015, a team at the Plymouth Marine Laboratory in England released film they had captured under a microscope showing zooplankton, tiny marine animals like krill, eat microplastic that has picked up a covering of biological material, such as algae, that mimics the smell of fish food. Research published findings in the journal *Science* found that juvenile perch actively preferred polystyrene particles to the plankton of their normal diet. One study by the Scripps Institute found that fish in the North Pacific ingest as much as 24,000 tons of plastic debris a year. And up the food chain it to goes to us, also increasing toxins along the way.

Chelsea Rochman, from the UC Davis School of Veterinary Medicine, and her team visited a fish market in Half Moon Bay in California. They sampled 76 fish from 12 species and one shellfish species. All had been caught nearby. The animals were dissected to reveal any plastic and fiber debris they contained. In total, 67 percent of the fish species contained plastic debris. Other species include:

- Sea turtles mistake plastic bags for jellyfish.
- Albatross and other sea birds feed bits of plastic to their young causing them to die of starvation and dehydration.
- Seals and other marine mammals often get caught in abandoned lost (ghost) fishing nets and debris.
- Filter feeders (like baleen whales) consume plastic bits instead of their normal plankton or fish eggs, which can clog their baleen "filters."
- Recent studies have revealed plastic pollution in 100 percent of marine turtles, 59 percent of whales, 36 percent of seals, and 40 percent of seabird species examined postmortem.
- Additionally, scientists from the National Marine Mammal Laboratory concluded that plastic entanglement was killing up to 40,000 seals a year.
- One million sea birds and 100,000 marine mammals are killed annually from plastic in our oceans. Forty-four percent of all seabird species, 22 percent of cetaceans, all sea turtle species, and a growing list of fish species have been documented with plastic in or on their bodies.
- Ocean Conservancy volunteers collected more than half a million straws and stirrers last year that are mistakenly eaten by sea turtles and seabirds. This causes their bellies to swell leading to starvation.
- Due to the increase in ocean plastic debris, by 2050, 99 percent of all species of seabirds will be eating plastic and 95 percent will fall victim to their harmful effects, up from 65 percent, a jump from the historical average of 26 percent.
- Of all the dolphins that get stranded and die on beaches, more than 70 percent have been found to have ingested some plastic debris.

- Experts from the Scientific Aspects of Marine Environmental Protection (GESAMP) states that plastic causes $13 billion of damage to the marine environment each year.

Origins of Ocean Pollution

Transportation vehicles (cars, trucks, trains, planes, and ships) emit carbon dioxide due to burning fossil fuels, leading to air pollution that reaches the ocean and becomes acid rain, which then pollutes the water, killing a variety of marine life.

Discharging sewage in the ocean has always been considered the cheapest, easiest, and most irresponsible way of disposing wastes. Roughly two-thirds of the world's marine life has been threatened with such compounds as household cleaners, medicines, plastics, and general garbage. And Alaskan forests are being befouled by tainted bear scat from polluted fish eaten by bears.

Oil spills suffocate marine life due to a breakdown in thermal insulation to those that survive. The gills of fish can be clogged by spilled or discharged oils, which can block off respiration. If sunlight is blocked by oil, marine plants will die because it affects photosynthesis. And this includes sunscreen and tanning oil.

Because of the cheap, crude, high-sulfur fuel used in merchant vessels they belch out almost one billion tons of carbon dioxide per year, the majority landing in the ocean. Sixteen of the world's largest ships upchuck more sulfur than all the cars on Earth, and a single large container ship can emit pollutants equivalent to 50 million cars.

Some 20 million people board cruise ships every year that flush about a billion gallons of sewage. Around 1.4 billion pounds of trash are dumped into the oceans each year by ships and boats, according to Friends of the Earth and the National Academy of Sciences.

Mining for precious minerals and oil in oceans, and the disposal of its wastes, is another source of pollution. In addition, with the increased traffic on the seas, the loud sounds from sonar devices and oil rigs are making noise pollution worrisome. According to researchers, this noise disrupts the migration and reproduction patterns of mammals like whales, dolphins, and other sea creatures.

The buildup of waste nutrients (mainly phosphorus, nitrogen, and carbon from fertilizers) from sources such as lawns and farmlands flow downriver to the sea and influences a lack of oxygen (hypoxia) in parts of the ocean called dead zones. They are invisible traps from which marine life can't escape, and the oxygen shortage causes the fish and other sea dwellers, such as lobsters and clams, to suffocate. In turn, it can sicken or kill people who eat contaminated seafood from the area.

Most scientists seemed to be stumped about what to do with the seagoing garbage dumps. For starters, only 1 percent of the marine litter floats, the vast majority sinks to the sea floor, and most of this plastic stew is made of pieces that are smaller than a grain of rice, which are simply unscoopable.

The Collapse of Coral Reefs

There was a time when scientists were cautiously optimistic that the oceans were absorbing immense amounts of carbon dioxide, supposedly relieving it from the atmosphere. Wrong again. The carbon dioxide our oceans absorb every day causes ocean acidification, for example, reducing the water's ability to carry calcium carbonate that corals need to build their skeletons.

It is estimated that by 2050, only 15 percent of coral reefs, which support more species and oxygen (per unit area) than any other marine environment, will have enough calcium carbonate for adequate growth, according to NOAA.

The coral reef structure buffers shorelines against waves, storms, and floods, helping to prevent loss of life, property damage, and erosion. They occupy less than 1 percent of the ocean floor, yet are home to more than a quarter of all marine species. Worldwide, coral reefs produce almost as much oxygen as all the forests, and there is 2 percent less oxygen in the ocean now than there was in 1950, according to the journal *Science.*

Coral reefs provide close to $30 billion each year in goods and services in the United States and form the nurseries for about a quarter of the ocean's fish, so it behooves us to keep them healthy.

Beach Bummers

Pollution on American beaches that violate public health standards happens thousands of times a year, mostly because of human and animal waste, and storm water runoff. The Environmental Protection Agency (EPA) estimates that up to 3.5 million people a year become sick from contact with contaminated beaches.

The National Resources Defense Council (NRDC) publishes an annual report called *Testing the Waters*, which tracks water quality at beaches around the country. The Beach Advisory and Closing Online (BEACON) and the EPA reports contain annual state-reported beach monitoring and notification data that is available online (www.epa.gov/.. ./beacon-20-beach-advisory-and-closing-online-notification).

The Center for Marine Conservation (CMC) reveals that among the debris found most frequently on beaches are cigarette butts, small plastic pieces (caps, lids, straws), food bags, glass and plastic bottles, Styrofoam cups, cans, and paper.

FRITTERING AWAY FRESH WATER

> No blue, no green, no life.—*Dr. Sylvia Earle*

It's odorless, colorless, tasteless, yet its physical caress is indescribable. It's quite widespread, yet sometimes hard to find. It's extremely precious, but in some places taken for granted. We can only survive for a few days without. It is the most wasted, and the most sought-after and fought-over element on Earth.

It's been on Earth since the planet was being formed, but its quality and useable quantity is deteriorating day by day now due to pollution and profligate waste. Conservation has become elemental for a myriad of reasons—from the epidemic occurrences of drought to climate change and it is fundamental, because it is the life blood of our biosphere.

Facts and figures:

- According to the EPA, more than 860 billion gallons of vile slurry slew, and eight trillion microbeads, escapes sewer systems across the country and pollutes the nearest body of water each year.
- In the United States there are 240,000 water main failures each year, and 10 times more from lesser municipal water pipes, wasting 1.7 trillion gallons. Two hundred million people could be

served by water lost to leaks, theft, and waste, according to the U.S. Geological Survey (GSA).

- Most of America's water pipes are 75 years old, and far too many sewer systems got rated a D– by American Water Works Association (AWWA), and the cost for replacement could reach more than $1 trillion.
- Montgomery County, a suburb of Washington, D.C., has the dubious honor of the leakiest place in the country—more than 4,000 water main leaks in four years.
- A leaky faucet can waste 100 gallons a day and costs the average household about $70 per year. If every household in America had a faucet that dripped once each second, 928 million gallons of water a day would leak away.
- The inattentive irrigation of U.S. lawns and landscapes alone claims an estimated 7.9 billion gallons of water a day, a volume that would fill 14 billion six-packs.
- Surface water—rivers and streams, lakes, reservoirs, and aquifers constitute approximately 80 percent of the water used on a daily basis.
- Of all water withdrawals, agricultural irrigation accounts for almost 40 percent—making it America's largest water-gorging sector.
- Public water supplies represent 11 percent of the total use.
- Less than 1 percent of all the water on Earth is potable.

A Waste of Fresh Water

Not all of Earth's fresh water sits on its surface. A great deal of water is held in underground structures known as aquifers. Back in 1996, a study in Iowa found that over half the state's groundwater was contaminated with weed killers and waste. In Nevada, due to atomic testing, an aquifer holding 1.6 trillion gallons of water is polluted with radiation.

Garbage and chemicals dumped legally and illegally by individuals and industries severely affect the water tables in our urban areas. Surface waste from rooftops, parking lots, and paved roads allow rain and snowmelt runoff to pick up heavy metals, oils, road salt, and other debris and contaminants that seep into aquifers.

Other pollutants include gross governmental blunders like the infamous situation in Flint, Michigan, where one mom stated, "Water coming out of the kitchen taps in our homes looked like frying oil, smelled like an open sewer, and contained lead."

Now one of the newest pollutants comes from our latest "frienemy," plastic waste. Scientists are finding microscopic plastic fibers in 83 percent of the world's drinking water and an astounding 94 percent in U.S. tap water, and in three leading brands of bottled water. The findings suggest that a person who drinks a liter (38.8 ounces) of bottled water a day might be consuming tens of thousands of microplastic particles each year. We already know that when microplastics build up in animals like fish, it affects their behavior and can alter their body chemistry. Some chemicals in plastic are known to have similar effects on humans.

These fibers escape into water every time we do our laundry, at a clip of more than 64,000 pounds a day. Too small to be captured by wastewater treatment, they then accumulate in our oceans and rivers. So far, there have been no governmental efforts to regulate products that contain these microfibers.

Untreated Waste in the Water

In America, 40 percent of the rivers and 46 percent of the lakes are considered unhealthy for swimming, fishing, or aquatic life. The EPA estimates that every year 1.2 trillion gallons of sewage from households, industry, and restaurants is dumped into U.S. waters. Bacteria, viruses, and parasites are released into recreational bodies of water mostly by bodily waste. But when they find their way into drinking water, some very serious and potentially lethal illnesses can result.

Some large contributors of organic waste are agricultural sources: vegetable and fruit packaging, processing of poultry and dairy, and meat packaging. Industry also produces pollutants from tanning, and the production of oil, paper, and wood.

Organic waste comes from a variety of sources, but the main source is domestic wastewater. We can take numerous actions to prevent water pollution from our homes. First, be attentive to throwing away trash and conscientious about recycling. Avoid lawn pesticides, collect pet poop and discard it responsibly, properly dispose of household chemicals and medicine, avoid products with microbeads, and attend to oil leaks on the mower or car. Also maintain septic tanks and inspect them every three to four years.

The Impediment of Inert Waste

Inorganic waste is also a problem. Any chemicals used in an industrial facility, such as cleaners, scrubbers, and acids, are sources of inorganic wastes. Also, salt and other ice-melting agents used on roads are very hard on marine animals and waterways. Water pollution through industrial waste is easily the most widespread type of water waste and pollution in the United States. Around 400,000 factories take clean water from rivers, and many pump polluted waters back in their place. Since 1970, the EPA has invested about $70 billion in improving water treatment plants. However, another $271 billion is still needed to upgrade the system.

In the past, liquid waste from industrial activities was dumped directly into rivers, or put into toxic waste barrels that were then buried or dumped in the oceans or waterways. The barrels deteriorated and leaked, resulting in heavily contaminated sites that are still being dealt with today.

In the past, rivers have actually caught on fire—the Rouge in Detroit, the Cuyahoga in Ohio, and the Buffalo River in New York, among others because of debris and flammable chemical waste. Those practices were abolished by the 1972 Clean Water Act, the Resource Conservation Recovery Act of 1976, and the Superfund Act of 1980.

What the Frack!

Fracking is a process that corporations use to extract oil and natural gas from rock formations deep underground by injecting a mix of water (billions of gallons), sand, and some very toxic chemicals at extreme pressure, to fracture the rock and release oil or gas.

Advocates insist it is a safe and economical source of clean energy. Critics claim it destroys drinking water supplies, polluting the air and land. The EPA released the final version of a six-year, $29 million study with the conclusion that hydraulic fracturing can cause contamination to drinking water resources.

The Wasted Energy of Water

It's estimated that U.S. water-related energy use is at least 521 million MWh (megawatt hours) a year—equivalent to a whopping 13 percent of the nation's electricity consumption, and up to a third of that is wasted. When water is wasted, more must be cleaned and transported, and this requires the use of more fossil fuels and other nonrenewable energy sources, which translates into more pollution.

Thirsty?

In the United States, drinking water is under threat from many forces, but none so consistently overlooked as aging, deteriorating infrastructure. Our nation's water systems are crumbling under the combined pressures of population growth, rapid urbanization, and chronic under-investment.

It is predicted that in fifteen years almost all states will be in a "water-stressed" condition. Yet water is still inexpensive enough that we don't think twice when wasting it. If drinking water and soft drinks were equally costly, your water bill would skyrocket more than 10,000 percent. You can refill an eight-ounce glass of water approximately 15,000 times for the same cost as a six-pack of soda.

Aggravating Agriculture

It is known that excessive irrigation can increase soil salinity and wash pollutants and sediment into freshwater ecosystems and species downstream. Much of the Southwest United States is desert, but about 90 percent of the Colorado River's water is diverted into these parched lands for agricultural irrigation. Perhaps half of this regional resource is wasted because it is lost to evaporation and seepage during pumping and transport, according to a Cornell University study in the journal BioScience.

Many farmers rely on flood irrigation, which is a highly inefficient means of delivering water to thirsty plants. The dwindling water flow of the Colorado River, the most litigated water source in the United States, threatens the supplies of California, Arizona, Nevada, Wyoming, Utah, Colorado, and New Mexico. Shaving irrigation water by 10 percent would save more than is used by individuals in all the seven states that it services.

There are other inefficiencies. Alfalfa, used as cattle fodder, is a relatively low-value crop but uses almost 80 percent of California's irrigation water because it is grown year-round in desert conditions. It contributes only 4 percent to the state's total farm revenue and pays only 15 percent of the capital costs of the federal system that delivers much of the irrigation water, according to the NRDC.

Water is especially needed in California, as the state contributes half the country's fruits, nuts, and veggies. Apologists claim that this is not a waste of water because it results in inexpensive produce for the country. No one wants to spend $10 for a head of lettuce. The agricultural value of the crops represents more than $105 billion in gross domestic product (GDP).

Willfully Wasted Water

Half of the world's population lives on 25 gallons of water or less daily. The average American uses between 100 and 175 gallons of water a day, wastes about half that amount, and believes that he or she uses only half as much as the actual amount, according to a survey done by Department of Public and Environmental Affairs at Indiana University.

The U.S. population doubled between 1950 and 2000, but demand for water rose 300 percent. Every day, leaking pipes waste an estimated seven billion gallons of clean drinking water (more than 11,000 swimming pools). That, combined with the $11 billion annual shortfall to replace aging water facilities, has made the United States a very water-inefficient country.

And things are going to get worse. By 2020, California estimates it will incur water shortages equal to the needs of 4 to 12 million families. A study released by the NRDC found that more than 1,100 U.S. counties face water scarcity as a result of climate change. Of those, 400 are in the "extreme risk" category, representing an increase of 14 times over previous shortage estimates.

Water Down the Drain

About 95 percent of the water entering a home goes down the drain—enough for more than 10 baths for every person on Earth—because many American households do not install water-efficient appliances.

According to the United States Geological Study, 97 percent of our fresh water is stored underground in aquifers that are being pumped dangerously low. The Ogallala Aquifer, one of the largest in the world, once held up to four quadrillion gallons of water. It yields about 30 percent of the nation's groundwater used for irrigation. The Ogallala Aquifer provides four trillion gallons every year, but some estimates have it containing only 20 percent of its original content by 2020 and possibly depleted by 2050, after only 150 years of human exploitation.

In Nevada, the water level at the Hoover Dam, supplied by the mighty Colorado River, has dropped 130 feet in the last 10 years. At this rate, the dam could be dry within the next decade.

Some Shortage Solutions

Toilet to Tap

High-tech efforts are in progress that some people might find rather unpalatable. Orange County, California, has spent more than $481 million on a system that takes human wastewater and turns it into drinkable water, colloquially called "toilet to tap." But, as far as the "ick" factor goes, when asked to identify their tap water from bottled water, people could not tell the difference. Its present capacity is 70 million gallons a day, and they are planning to increase that by 30 percent.

A Salty Solution

California's drought is never ending, and one of its answers is in the Pacific Ocean. San Diego, Huntington Beach, Santa Barbara, and maybe Marin County, all have plunged into the ocean to mitigate their water problems. The solution is a costly one (around $1 billion per plant) and it doesn't completely solve the problem, but ask a drought-plagued person what they are willing to pay for water. The tide is also turning in places like Florida and other coastal states.

Desalinization is not practical for most of the United States as most of the country has no access to seawater. Conservation is still the main watchword, and experts agree that it's the first and best course of action for managing freshwater.

Home H_2O Fixes

Ten percent of American homes squander 90 gallons of water on a daily basis. Leaks are costly. If a leak has a continuous flow of one-tenth of a gallon per minute, it wastes 180 gallons per day. And according to the EPA, more than 3,000 gallons per year can be lost from a leaky faucet that drips at the rate of one drip per second. The kitchen accounts for about 20 percent of our indoor water use. Letting your faucet run for just five minutes while washing dishes can waste up to 25 gallons of water and uses enough energy to power a 60-watt light bulb for 18 hours. If people get more water wise, just brushing their teeth with the water turned off could save more than four times the Mississippi River's annual flow of water (593,003 cubic feet per second) in one year.

More than 47 percent of water use in the average American home occurs in the bathroom, with nearly 35 percent being flushed, 40 percent used in showers and baths, 15 percent from faucets, and 10 percent for leaks.

Older toilets can use three gallons of clean water with every flush, while ultra-low flush (ULF) toilets use as little as one gallon. A leaky toilet can waste more than 10,000 gallons of water a year. The average family can save 13,000 gallons of water and $130 to $330 in water costs per year by replacing their old, inefficient toilets with WaterSense-labeled models. If just 1 percent of households replaced high-flush toilets with products approved by the EPA WaterSense program, the nation could save 520 billion gallons of water per year, wouldn't waste approximately $38 billion of electricity, and would prevent 80,000 tons of greenhouse gas emissions—equivalent to removing nearly 15,000 automobiles from the road for one year.

If you shower longer than five minutes, you're wasting five to 10 gallons of water every extra minute, according to EarthEasy.com. Reducing your shower time by just one to two minutes can save up to 750 gallons of water a month, according to Water Use It Wisely. Using a low-flow showerhead can save up to 800 gallons of water per month. If you have to take a bath, do it with friends and family to save water and energy, and try to shower the navy way—rinse, lather, rinse, and out.

Clothes washers are responsible for about 20 percent of total water use and are the second-largest consumer of water in most homes. Running half-full loads of laundry can waste 1,000 gallons of water a month. The EPA states that a standard washing machine uses 41 gallons of water per load, but a high-efficiency washing machine uses less than 28 gallons of water, saving 6,000 gallons of water per year for an average family front-load washer. In addition, concentrated detergents could save billions of gallons of water because they work efficiently and avoid the expense of manufacturers adding water to their products and spending money and energy to ship it.

Nearly 60 percent of a household's water footprint can go toward lawn and garden maintenance, and more than 50 percent of landscape water is wasted due to overwatering, inefficient watering practices, and broken or poorly maintained irrigation systems. Using soaker hoses or drip irrigation in flower beds can save up to 50 percent of the water used compared to sprinklers, which can consume up to 265 gallons an hour. Use sprinkler heads either in the early morning hours or at dusk to avoid evaporation. Watering on windy days gets water blown away and evaporates faster.

Wash the car (not the driveway) and be sure to turn off the nozzle when not in use. That will save you as much as 70 gallons per wash. A home car wash will consume 80 to 140 gallons of water, whereas an efficient carwash will use about half that much, and their water is recycled.

A swimming pool naturally loses 1,000 gallons or more a month to evaporation and water play alone. More is lost from cracks in pools—up to a whopping 20 to 30 percent per year. Using a pool or hot tub cover reduces water loss, and the hotter the water in your spa or pool the more it evaporates—so keep the heat setting moderate.

Use a broom to clean sidewalks instead of a hose. A homeowner who neglects to regularly replace rubber gaskets in outdoor faucets can waste hundreds of gallons of water and increase water bills.

How many gallons a day do you use? Check your water bill.

- 80 gallons per day—Excellent. You use water wisely. Please share your conservation techniques with friends and neighbors.

- 80 to 100 gallons per day—Very good. You use less water than the average citizen.
- More than 120 gallons per day—Water hog! Get some conservation tips and learn how you can conserve water.

For more information see my books on conservation at home and at work: *The Energy Wise Home* and *The Energy Wise Workplace*, published by Rowman & Littlefield.

SOURCES

"10 Easy Ways to Save Water in Your Yard and Garden." Love Your Landscape.org, https://www.loveyourlandscape.org/expert-advice/water-smart-landscaping/water-saving-tips/10-easy-ways-to-save-water-in-your-yard-and-garden/.

"40 Interesting Facts about Water Pollution." Conserve Energy Future, https://www.conserve-energy-future.com/various-water-pollution-facts.php.

"Ag and Food Sectors and the Economy." U.S. Department of Agriculture, https://www.ers.usda.gov/data-products/ag-and-food-statistics-charting-the-essentials/ag-and-food-sectors-and-the-economy.aspx.

Almeda, Rodrigo, et al. "Interactions between Zooplankton and Crude Oil: Toxic Effects and Bioaccumulation of Polycyclic Aromatic Hydrocarbons." *PlosOne*, June 28, 2013, http://journals.plos.org/plosone/article?id=10.1371/journal.pone.0067212.

"California Agricultural Production Statistics." California Department of Food and Agriculture, February 7, 2018, https://www.cdfa.ca.gov/statistics/.

Canales, Kylie. "Water: An Introduction to Industrial Waste." *CHEnected*, September 16, 2011, https://www.aiche.org/chenected/2011/09/water-introduction-industrial-waste.

"Causes and Effects of Water Pollution." Go Green Academy, http://www.gogreenacademy.com/causes-and-effects-of-water-pollution/.

"Causes of Water Pollution." EcoAmbassador, https://www.theecoambassador.com/Causesofwaterpollution.html.

Cho, Renee. "Losing Our Coral Reefs." *State of the Planet* (blog), Earth Institute, Columbia University, June 13, 2011, http://blogs.ei.columbia.edu/2011/06/13/losing-our-coral-reefs.

Chua, Celestine. "Our Culture of Waste: Why We Should Stop Wasting (and How to Prevent Waste), *Personal Excellence* (blog), https://personalexcellence.co/blog/how-to-prevent-waste/

Clean Water Act, Wikipedia, https://en.wikipedia.org/wiki/Clean_Water_Act.

Clift, Jon, and Amanda Green Cuthbert. "How Much Water Do You Use? Here's Some Quick Numbers." *Alternet*, August 4, 2009, https://www.alternet.org/story/141751/how_much_water_do_you_use_here%27s_some_quick_numbers.

"Conserve Water to Preserve Life on Earth (and save money)." *Care2* Petitions, https://www.thepetitionsite.com/764/971/651/conserve-water-to-preserve-life-on-earth-and-save-money/.

Dondero, Jeff. *The Energy Wise Workplace*. Lanham, MD: Rowman & Littlefield, 2017.

Dondero, Jeff. *The Energy Wise Home* Lanham, MD: Rowman & Littlefield, 2017.

"Drip Calculator: How Much Water Does a Leaking Faucet Waste?" The U.S. Geological Survey, December 2016, https://water.usgs.gov/edu/activity-drip.html.

Duffy, Tammy. "Beach Contamination: A Hurricane Reality." *Duffy's Cultural Couture* (blog), October 22, 2016, http://www.tammyduffy.com/ARTFASHION/index.blog?from=20161022.

"Flow of Colorado River in Seven Downstream States Threatened." *Denver Post*, January 5, 2018, https://www.denverpost.com/2018/01/05/colorado-river-snowpack-threatens-flow-downstream/.

Frabricius, Karl. "The North Pacific Gyre: 100 Million Tons of Garbage and Growing." *Scribol*, August 18, 2009, http://scribol.com/environment/waste-and-recycling/the-north-pacific-gyre-100-million-tons-of-garbage-and-growing/.

Gastaldo, Evann. "Water in Flint, Michigan, Looks 'Like Urine,' and Worse." *Newser*, October 2, 2015, http://www.newser.com/story/213863/water-in-flint-michigan-looks-like-urine-and-worse.html.

Geer, Abigail. "Ways Water Pollution Is Killing Animals." *Care2*, August 20, 2017, https://www.care2.com/causes/5-ways-water-pollution-is-killing-animals.html.

Geib, Claudia. "US Absent from Global Meeting on Ocean Health. That's Bad News for the Planet." *Futurism*, March 12, 2018, https://futurism.com/us-absent-global-meeting-ocean-health/.

Gertz, Emily J. "Ocean Plastic Pollution's Shocking Death Toll on Endangered Animals." *TakePart*, February 20, 2015, http://www.takepart.com/article/2015/02/20/ocean-plastic-pollutions-shocking-death-toll-endangered-animals.

"Great Pacific Garbage Patch." Wikipedia, https://en.wikipedia.org/wiki/Great_Pacific_garbage_patch.

Griffiths-Sattenspiel, Bevan. *Water-Energy Toolkit: Understanding the Carbon Footprint of Your Water Use*. Boulder, CO: River Network, 2010, https://www.rivernetwork.org/wp-content/uploads/2015/10/Toolkit_Emissions2-8-12.pdf.

Guo, Jeff. "Agriculture Is 80 Percent of Water Use in California. Why Aren't Farmers Being Forced to Cut Back?" *Washington Post*, April 3, 2015, https://www.washingtonpost.com/blogs/govbeat/wp/2015/04/03/agriculture-is-80-percent-of-water-use-in-california-why-arent-farmers-being-forced-to-cut-back/?utm_term=.ed076f7fb84.

Hadhazy, Adam. "Top 10 Water Wasters: From Washing Dishes to Watering the Desert." *Scientific American*, July 23, 2008, https://www.scientificamerican.com/article/top-10-water-wasters/.

Hertel, Lauren. "Recovery from Ocean Warming Can Take Thousands of Years." The Switzer Foundation, April 6, 2015, https://www.switzernetwork.org/leadership-story/recovery-ocean-warming-can-take-thousands-years.

Hinton, Kevin. "The Effects of Ocean Dumping." Environment 911, http://www.environment911.org/The_Effects_of_Ocean_Dumping.

Honeywell, John. "What Happens When You Flush the Loo on a Cruise Ship? *Telegraph*, February 1, 2017, https://www.telegraph.co.uk/travel/cruises/what-happens-when-you-flush-the-loo-on-a-cruise-ship-.

"How Do Coral Reefs Protect Lives and Property?" National Ocean Service, https://oceanservice.noaa.gov/facts/coral_protect.html.

"How Much Water Do You Use in a Shower?" *Constellation* (blog), https://blog.constellation.com/2016/07/05/average-shower-length-flowchart/.

"How Much Water Does a Sprinkler Use Per Hour?" *SFGate*, http://homeguides.sfgate.com/much-water-sprinkler-use-per-hour-82122.html.

Huynh, Christina. "Five Ways to Decrease Your Kitchen Water Use." *Green Home Guide*, July 7, 2016, http://www.greenhomeguide.com/know-how/article/five-ways-to-decrease-your-kitchen-water-use.

"Impacts and Carrying Capacity: Environmental Impacts from Unsustainable Population Growth." World Population Awareness, World Overpopulation Awareness, http://www.overpopulation.org/impact.html.

Inskeep, Benjamin D., and Shahzeen Z. Attari. "The Water Short List: The Most Effective Actions U.S. Households Can Take to Curb Water Use." *Environment*, July–August 2014, http://www.environmentmagazine.org/Archives/Back%20Issues/2014/July-August%202014/water_full.html.

James, Adam. "The U.S. Wastes 7 Billion Gallons of Drinking Water a Day: Can Innovation Help Solve the Problem?" *ThinkProgress*, November 3, 2011, https://thinkprogress.org/

the-u-s-wastes-7-billion-gallons-of-drinking-water-a-day-can-innovation-help-solve-the-problem-f7877d6e3574/.

Kirkpatrick, Noel. "Fish Ingesting Plastic from Great Pacific Garbage Patch." *Mother Nature Network*, April 2018, https://www.mnn.com/earth-matters/animals/stories/fish-ingesting-plastic-from-great-pacific-garbage-patch.

Kurth, Jim. "Marine Debris and Wildlife: Impacts, Sources, and Solutions." Statement to Congress, U.S. Department of the Interior, May 17, 2016, https://www.doi.gov/ocl/marine-debris.

Lance, Jennifer, "Toilet to Tap: Orange County Turning Sewage Water into Drinking Water." *Insteading* (blog) (2011), https://insteading.com/blog/toilet-to-tap-orange-county-turning-sewage-water-into-drinking-water/.

"Learn about the Effects of Pollution on Freshwater." *National Geographic*, https://www.nationalgeographic.com/environment/freshwater/pollution/.

Le Guern, Claire. "When The Mermaids Cry: The Great Plastic Tide." *Coastal Care*, March 2018, http://coastalcare.org/2009/11/plastic-pollution.

Leverette, Mary Marlowe. "How to Save Water in the Laundry Room." *The Spruce*, October 3, 2017, https://www.thespruce.com/save-water-in-the-laundry-room-2146003.

"Litter—Pervading the Oceans." *World Ocean View* (2010), https://worldoceanreview.com/en/wor-1/pollution/litter/.

McGrath, Matt. "Fish Eat Plastic Like Teens Eat Fast Food." *BBC News*, June 2016, http://www.bbc.com/news/science-environment-36435288.

Mills, Richard. "Water: An Endangered Global Resource." *Ahead of the Herd*, http://aheadoftheherd.com/Newsletter/2012/Water-An-Endangered-Global-Resource.htm.

Misachi, John. "Rivers That Have Caught on Fire." *World Atlas*, April 25, 2017, https://www.worldatlas.com/articles/is-the-cuyahoga-river-the-only-river-to-ever-catch-on-fire.html.

"The Plastic Plague: Can Our Oceans Be Saved from Environmental Ruin?" CNN, September 2, 2106, https://www.cnn.com/2016/06/30/world/plastic-plague-oceans/index.html.

Montgomery, David. "Drought Returns to Huge Swaths of US, Fueling Fears of a Thirsty Future." *Stateline*, April 17, 2018, http://www.pewtrusts.org/en/research-and-analysis/blogs/stateline/2018/04/17/drought-returns-to-huge-swaths-of-us-fueling-fears-of-a-thirsty-future.

Mosbergen, Dominique. "The Oceans Are Drowning in Plastic—and No One's Paying Attention." *Huffington Post*, May 12, 2017, https://www.huffingtonpost.com/entry/plastic-waste-oceans_us_58fed37be4b0c46f0781d426.

"New Clothes Washer and Dishwasher Standards Will Save Consumers Loads of Money, Protect the Environment." American Council for an Energy Efficient Economy, May 16, 2012, http://aceee.org/press/2012/05/new-clothes-washer-and-dishwasher-st.

Oskin, Becky. "What Is Groundwater?" *Live Science*, January 8, 2015, https://www.livescience.com/39579-groundwater.html.

Pearce, Fred. "How 16 Ships Create as Much Pollution as All the Cars in the World." *Daily Mail*, November 29, 2009, http://www.dailymail.co.uk/sciencetech/article-1229857/How-16-ships-create-pollution-cars-world.html.

"Plastic Fibres Found in Tap Water around the World, Study Reveals." *Guardian*, September 6, 2017, https://www.theguardian.com/environment/2017/sep/06/plastic-fibres-found-tap-water-around-world-study-reveals.

Reddy, Puli Naveen. "What Is Not a Crime Today but Will Be in 50 Years?" *Quora*, December 13, 2016, https://www.quora.com/What-is-not-a-crime-today-but-will-be-in-50-years.

Reynolds, Kelly A. "Gone Water Gone." *The Aquifer* 28, no. 4 (Spring 2014), http://www.groundwater.org/file_download/inline/138784d1-7e55-4cfe-98e9-f3ad29ec8e64.

Rinkesh. "Causes and Effects of Ocean Dead Zones." *Paperblog*, September 2016, https://en.paperblog.com/causes-and-effects-of-ocean-dead-zones-1536993/.

Rockwell, Susanne. "9 Ways UC Davis Is Rescuing Oceans." UC Davis, June 6, 2016, https://www.ucdavis.edu/news/9-ways-uc-davis-rescuing-oceans.

Rogers, Paul. "California Water: Desalination Projects Move Forward with New State Funding." *San Jose Mercury News*, January 29, 2018, https://www.mercurynews.com/2018/01/29/california-water-desalination-projects-move-forward-with-new-state-funding/.

Sabol, Colin. "The State of Water in America." *State of the Planet* (blog), Earth Institute, Columbia University, March 22, 2011, http://blogs.ei.columbia.edu/2011/03/22/water-in-america-2/.

Scheck, Tom, and Scott Tong. "EPA Reverses Course, Highlights Fracking Contamination of Drinking Water." *APM Reports*, December 13, 2016, https://www.apmreports.org/story/2016/12/13/epa-fracking-contamination-drinking-water.

Shaban, Bigad. "Microbeads Could Be Harming the Environment, Scientists Say." *CBS News*, May 6, 2014, https://www.cbsnews.com/news/microbeads-could-be-harming-the-environment-scientists-say/.

Smillie, Susan. "From Sea to Plate: How Plastic Got into Our Fish." *Guardian*, February 14, 2017, https://www.theguardian.com/lifeandstyle/2017/feb/14/sea-to-plate-plastic-got-into-fish.

"Statistics and Facts." U.S. Environmental Protection Agency, https://www.epa.gov/watersense/statistics-and-facts.

Stern, Denise. "What Are Different Ways That People Waste Water?" *Livestrong*, June 13, 2017, https://www.livestrong.com/article/119276-different-people-waste-water/.

"Surface Runoff." WikiVisually, from Wikipedia, https://wikivisually.com/wiki/Surface_runoff.

"Ten Things You Can Do to Reduce Water Pollution." Town of Simsbury, Connecticut, https://www.simsbury-ct.gov/water-pollution-control/pages/ten-things-you-can-do-to-reduce-water-pollution

"The 10 Worst Industrial Causes of Water Pollution (A Man-Made Crisis?)." All About Water Filters, http://all-about-water-filters.com/worst-industrial-causes-of-water-pollution/.

"Thousands of Miles Away Is Not Far Enough to Escape Plastic Pollution." Blue Ocean Network, https://blueocean.net/thousands-miles-away-not-far-enough-escape-plastic-pollution/.

"Threats Facing the Marine Environment: Marine Noise Pollution." *Marine Life*, http://www.marine-life.org.uk/conservation/threats-facing-the-marine-environment/marine-noise-pollution.

"Toilet." Conserve h2o.org. https://www.conserveh2o.org/toilet-water-use.

"Toilets, Frugal Flushing and Water Saving Tips." Home Water Works, https://www.home-water-works.org/indoor-use/toilets.

Vartabedian, Ralph. "Nevada's Hidden Ocean of Radiation." *Los Angeles Times*, November 13, 2009, http://articles.latimes.com/2009/nov/13/nation/na-radiation-nevada13.

"Water." Scribd, https://www.scribd.com/doc/40438169/Water.

"Water and Waste Water Systems." Hernando County Utilities Department, http://www.hernandocounty.us/utils/Water_Sewer/facts.htm.

"Water Conservation and Washing Vehicles." Maryland Department of the Environment, http://mde.maryland.gov/programs/water/waterconservation/Pages/carwashing.aspx.

"Water Consumption by the United States." https://public.wsu.edu/~mreed/380American%20Consumption.htm.

"Water Facts." UN Water, http://www.unwater.org/water-facts/5/?ipp=10&tx_dynalist_pi1%5Bpar%5D=YToxOntzOjE6IkwiO3M6MDoiIjt9.

WaterSense. *Water-Smart Landscapes*. Washington, DC: U.S. Environmental Protection Agency, 2017, https://www.epa.gov/sites/production/files/2017-01/documents/ws-outdoor-water-efficient-landscaping.pdf.

Weule, Genelle. "Plastic and How It Affects Our Oceans." *Science*, February 27, 2017, http://www.abc.net.au/news/science/2017-02-27/plastic-and-plastic-waste-explained/8301316.

"What Is Fracking?" Food and Water Watch, https://www.foodandwaterwatch.org/problems/fracking.

"When In Drought—Tips for Reducing Water Usage." Davidson Communities (blog), http://www.davidsoncommunities.com/2015/10/when-in-drought-tips-for-reducing-water-usage/.

Woodford, Chris. "Water Pollution: An Introduction." *Explain That Stuff*, June 4, 2017, http://www.explainthatstuff.com/waterpollution.html.

Yator, Geofrey. “Agriculture and Its Effect on the Environment.” *The Environment*(blog), November 15, 2012, http://theenviro.blogspot.com/2012/11/agriculture-and-its-effect-on.html

Yin, Sandra. “Lifestyle Choices Affect U.S. Impact on the Environment.” *PRB*, October 2006, https://www.prb.org/lifestylechoicesaffectusimpactontheenvironment.

“Your Cool Facts and Tips on Waste Management.” EschoolToday, http://www.eschooltoday.com/waste-recycling/sources-of-waste.html.

Zielinski, Sarah. “The Colorado River Runs Dry.” *Smithsonian Magazine*, October 2010, https://www.smithsonianmag.com/science-nature/the-colorado-river-runs-dry-61427169/#teMbqy2lPOcAjjCh.99.

4

LOSING LAND

A nation that destroys its soil destroys itself.
—*FDR*

THE ECOSYSTEM ANCHOR

Land is literally the platform and anchor of our biosphere. It's the secure ground of home—where we live, walk, build, and grow our food—and where we also pollute, waste, and destroy. While we can see plastic beaches, air pollution, and oil slicks, you can't see poisons seeping into the soil slowly turning it into noxious, unusable dirt. So far, the Environmental Protection Agency (EPA) has identified 1,408 hazardous waste sites as the most serious in the nation—all on land.

Land pollution is the introduction into the environment of substances that don't normally belong there and can cause long-term damage, destruction, degradation, or loss, as well as harmful effects on plants, animals, and humans.

Land takes up about a third of the planet's surface. We are over seven billion strong trying to survive here, and our lives are intimately tied to it. Increasing population and land developments bring landfills and reclamations and leave waste behind, leading to further deterioration of land. In addition, due to the lack of green cover, soil erosion and landslides occur, washing away fertile portions of the land. The increasing numbers of barren land plots and the decreasing acreage of forest cover is alarming.

Not even the most isolated places can be considered completely safe from pollution, even if they are hundreds or thousands of miles from the nearest human settlement, even if no human has ever lived there. A good example is Antarctica, where only small bands of scientists spend time in isolated bases, yet core samples of ice show signs of pollution.

The loss of quality and/or productivity of land as a place for agriculture, forestation, development, and living is often anthropogenic (human-made) in origin. In many places, land loss is due to what we call natural causes, like storms, rising tides, natural wildfires, and innate and inherent changes in the environment. But some of these changes, like the number and severity of storms and rising seas, are also anthropogenic, caused by pollution and overdevelopment of built environments, which are needed, but so is sensible direction and self-control. What gets us in trouble is when economic arguments try to justify ecological destruction.

Facts and figures:

- The United States loses topsoil 17 times faster than it's being formed, and 500 years are needed for the regeneration of one inch (2.5 cm) of topsoil.
- According to the U.S. Fire Administration, thousands of fires a year start due to natural ignition of landfills because of the heat generated from rotting material.
- Twenty-two barrels of toxic waste were disposed in the Love Canal, in the late 1970s. Many of the families living nearby later suffered from leukemia and high red blood cell counts. Forty years later it is still pulsating poison and residents are suing again claiming the Love Canal was never properly remediated and dangerous toxins continue to leach onto peoples' properties.
- Greenpoint oil spill is the biggest that has ever occurred in the United States. An estimate of between 17 and 30 million gallons of oil spilled and later leaked into the soil causing one of the biggest land pollution disasters in U.S. history. The spill has been oozing under and around Greenpoint for five decades, destroying the local aquifer, and rendering hundreds of acres of land severely compromised.

- The planet loses 24 billion tons of topsoil each year because of land pollution.
- Agricultural activities, anything related to fertilizing, using pesticides, herbicides, irrigation and improper disposal of animal waste, are the reasons for a significant percent of the land pollution.
- Worms absorb most of the toxins in the soil, which are then transferred up the food chain.
- Pollution of the land by chemical spills can be devastating and irreversible. The 100 mile swath of devastation downwind from the International Nickel refinery in Sudbury, Ontario, that was destroyed by pollution has not recovered, though pollution emissions dropped by a factor of five after the chemical spill. Once there were farms and forests; now there is nothing but bare rock.
- Research has calculated that nearly 33 percent of the world's adequate or high-quality food-producing land has been lost to erosion or pollution at a rate that far outstrips the pace of natural processes to replace diminished soil.

LANDFILLS—MORE THAN MEETS THE EYE AND NOSE

The most common and ancient method of discarding all sorts of rubbish was simply to bury it—out of sight out of mind. But that could only last so long. Soon enough, it became a source of contaminants that destroyed all the soil nearby, and "dumps" became simply holes in the ground where trash was buried. They offered no environmental protection, were not regulated, were an attraction to vermin, and could cause and spread disease.

Landfills, on the other hand, are carefully designed structures built into or on top of the ground, in which trash is buried or separated. When constructed carefully, they contain garbage and serve to prevent contamination between the waste and the surrounding environment, especially groundwater.

Unfortunately, they are not designed to biodegrade trash, merely to keep it buried. Because landfills contain minimal amounts of oxygen and moisture, trash does not break down rapidly.

Landfills are referred to as engineered, state-permitted disposal facilities where municipal solid waste (MSW; nonhazardous waste generated from family residences, hotels, and commercial and industrial waste) may be disposed of for long-term care and monitoring. All modern MSW landfills must meet regulations to ensure environmentally safe and secure inert disposal facilities.

The top and bottom liners, made of durable, puncture-resistant synthetic high-density polyethylene (plastic), separates and prevents the buried waste from coming in contact with the native soils and groundwater.

"Cells" are the areas in a landfill that have been constructed and approved for disposal of waste. The cells range in size from a few acres to 20 acres or more. The waste coming into the landfill for disposal is prepared by placing the material in layers where it is then compacted and shredded by heavy landfill machinery. Then it is covered daily with either six inches of compacted soil, or an alternative cover such as a spray-on foam or flame-retardant fiber material, or a tarpaulin-type fabric and is laid over the waste each day and removed the next day before more rubbish is placed on top.

The bottom of each landfill is typically designed at a slope where any liquids will drain and be trapped inside the landfill (the sump). There leachate (fouled water) is collected and removed from a layer of sand or gravel placed in the bottom of the landfill by a series of perforated pipes and pumped or moved by gravity to a holding tank or pond, where it is either treated on-site or hauled off-site to a public or private wastewater treatment facility. Groundwater monitoring stations are set up to access and test the groundwater around the landfill for presence of leachate chemicals.

During rain, the runoff is directed through a series of berms or ditches to holding areas known as sedimentary ponds. The runoff water is held long enough in these ponds to allow the suspended soil particles to settle before the water is pumped out.

In the absence of oxygen, bacteria in the landfill break down the trash. This process produces methane gas. Since methane is flammable, it has to be removed by a series of pipes that are embedded within the landfill. Once collected, it can be naturally vented, burned off, or used as an energy source.

LAND POLLUTION PRODUCTION

Caustic Coal

Ash from coal-fired plants contains chemicals so poisonous—having more radioactivity than nuclear power plant waste—that it is considered carcinogenic and can cause development of reproductive disorders in people. Most of the metals in coal ash also contain lead PAHs (polynuclear aromatic hydrocarbons). If the concentration of PAHs in the soil is high, the risk of potential dangers increases vastly. Exposure to PAHs may cause harmful health effects, and has caused cancer in animals, according to the EPA.

When various pollutants that circulate in the air get mixed with falling rain, it can transform into acid rain. The structure of the soil and the nutrients inside are very sensitive to the acidic substances and can get damaged significantly or destroyed by it.

Mining

The process of extracting minerals and other geological materials requires methods of operation that are not what would be called eco-friendly. For example, mining goes hand-in-glove with deforestation because space is required to mine, and trees obviously get in the way. Also, mining removes valuable native topsoil and organic matter that is difficult to reproduce.

One ton of copper produces two tons of waste, much of which is "tailings," a mud-like effluent that has been described as a major environmental problem—the largest environmental liability of the mining industry. Tailings are usually highly toxic and a huge percent of the waste gets dumped at a close and convenient landfill, creating a cycle of land pollution that is difficult if not impossible to rectify. Tailings containment facilities are regarded as the world's largest human-made objects. Oil sands tailings are the residue that remains after bitumen (petroleum resource) is separated from soil, and when dumped it can cause the land to be unusable for anything else. They create very long-term environmental liabilities that future generations will have to manage.

Historically, the alternative to storage was to dispose of tailings in the most convenient way possible (such as river dumping), which led to widespread environmental contamination in mining areas. Most modern hard-rock metal mines today dispose of tailings in pits lined with clay or a synthetic liner, or they are deposited back into the original mining pit. Some large mines seal off entire valleys with earthen dams, and others store tailings in natural lakes. In most cases, disposal pits are covered with water, forming an artificial lake, which reduces the rate of acid formation.

The environmental impact of mining includes erosion, formation of sinkholes, loss of biodiversity, and contamination of soil, wildlife, groundwater, and surface water by chemicals. However, most mining in the United States is strictly controlled.

Overgrazing

As land is overgrazed, native plants decline and foreigners such as cheatgrass, a tough, wild, invasive grass, can take over. It's not good for grazing, is a fire hazard that can destroy native juniper and pine forests, thus opening more room for cheatgrass and further reducing land for livestock ranching.

Deforestation

Trees are an integral part of the life of all living species. They provide natural habitat for animals, provide oxygen, remove CO_2, supply wood resources, and replenish the soil with nutrients that reproduce trees and plant life. But half the world's rainforests have been razed in the last century. At current rates, they will vanish altogether in the next century. Deforestation interrupts natural cycles and results in altering an eco-system forever.

The consensus of scientists is that up to 15 percent of all human climate emissions, nearly as much as one billion vehicles, now comes from deforestation, mostly in tropical areas. It can affect rainfall and weather worldwide, according to *Scientific American*.

If all countries stick to their pledges to decrease deforestation and let damaged forests recover, as promised at the Paris Accord in 2015, annual global greenhouse gas emissions could be reduced by as much as 24 to 30 percent, an enormous step. Wealthy countries pledged to raise $100 billion a year to help poor countries reduce their emissions. Some of that money will go to tropical forest protection. Unfortunately, Donald Trump has pulled the United States out, breaking our Paris Accord promises, claiming that the lowering of climate change by a "tiny, tiny amount," is not worth the job loss in the United States.

Every year, an area of rainforest the size of England and Wales is cut down (that's about 57 percent the size of California or 263,442 square miles). Unless something is done, they will vanish altogether in the next 100 years, according to *Earth Observatory*.

The Amazon is the largest rainforest on Earth; spanning nine countries, covering 2.6 million square miles and providing habitats for a staggering 10 percent of all known plant and animal species, and is the source of 20 percent of all fresh water worldwide.

An analysis of satellite images of the Brazilian part of the Amazon basin, which forms part of the largest contiguous rainforest on Earth, shows that on average 6,000 square miles of forest is being cut down by selective logging each year. Another 4,800 square miles is clear-cut annually, or double the rate of all previous estimates, for minerals, timber, cattle grazing or farming, according to the *Guardian*. Eighty percent of all timber exported from the Amazon in the 1990s was found to have been illegally harvested.

The faster trees go, the slimmer the chance of reversing climate change. The point of no return for the Amazon, which would result in a major ecological collapse, has been estimated at 20 percent deforestation. We are currently at 18 percent.

The problem is that land is stationary, so land pollution stays where it is until and unless someone cleans it up. That means deforestation, landfill sites, and radioactive waste dumps indefinitely take land out of circulation. Simple math tells us that the more land we use up, the less we have remaining to use.

Farming

The demand for food has increased with the population, and its ecological footprint is growing larger. New York's ecological footprint (the amount of land needed to support it) is 125 times bigger than the city itself.

The boon of technology has become a boondoggle for nature. Farmers often use highly toxic fertilizers and pesticides to get rid of insects, fungi, and bacteria from their crops. However, with the overuse of these chemicals, the result is contamination and poisoning of the soil. Biosolids, also known as sewage sludge, have been used as fertilizers in the past. However, recent studies and statistics show that such substances are more likely to be harmful to the ground because of the contaminated chemicals they contain. Up to 98 percent of herbicides and 95 percent of pesticides end up going deeper than the topsoil. That interferes with the biodiversity in the ground because of the reduction of organic matter and can contaminate the water table and the soil.

When crops are harvested, some pollutants can find their way into the food chain and affect the animals and people eating them—a good reason why we should be concerned about the condition of the land in which we plant and grow.

Soil Erosion

Soil is far more complex than just "dirt." Too much wind or water, excessive planting, overkill of nutrients and fertilizer, and overgrazing all cause soil degradation, damaging its structure and drastically reducing its productivity, until it's little more than dead dirt. Although nature can have a hand in this, human activities are the leading reason for the drastic increase in this phenomenon. Human interferences include building roads, intensive agricultural use, deforestation, and climate change. It all contributes to the wasting away of soil.

By the way, scientists have identified 414 towns and cities in the United States that are guaranteed to eventually be underwater, regardless of how much we decrease our carbon emissions, according to Climate Central, the group who conducted and published the study in the *Proceedings of the National Academy of Sciences* journal. Is your city on the doomsday list? (see http://www.dailymail.co.uk/sciencetech/article-3274955/Will-city-survive-century-Interactive-map-reveals-414-doomed-cities-obliterated-rising-sea-levels.html).

Urbanization

Humans have been making permanent settlements for at least 10,000 years and will hopefully continue into the future. Surprisingly, all the land mass of cities only comes to about 2 to 3 percent of the planet, yet it contains close to 60 percent of the Earth's population. People have to live and work somewhere, but urbanization changes the landscape and causes land to change in a variety of ways. One of the obvious problems of urban areas is that along with more people come their built environments, altering or obliterating ecosystems, and creating lots of waste.

Road Rubbish

The corridor of land on either side of roads and highways systematically becomes wasted with time and all kinds of harmful by-products of road travel—fuel spills, brake linings, tire trash, litter, grime from pavement, salt, and heavy-metal deposits shed from engines. These chemicals accumulate in the soil and combine to form a toxic stew.

WILDLIFE WORRIES

The animal kingdom faces a serious threat with regard to the loss of habitat. It occurs when enough change happens in an area so that it can no longer support natural wildlife. Degradation is when a habitat becomes polluted or is invaded by a nonnative species, crowding out or killing original plants and animals. Although most of the waste we produce is relatively harmless and easy to dispose of, around one-fifth of it is dangerous or toxic and extremely difficult to get rid of without automatically contaminating land.

Increasing food production is a major agent for the conversion of natural habitat into agricultural land. The constant human activities that leave land altered or polluted force natural species to move farther away, adapt to new regions, or die trying.

HOPE FOR THE LAND

Besides moderating our production, and increasing recycling and reuse, we have to stop befouling the land and clean up contaminated sites that already exist. Where sites can't be completely restored, it's possible to "reuse" them in other ways, for example, by turning them into wind or solar farms.

Bioremediation is a promising land-cleaning technology, in which microbes of various kinds eat and digest waste and turn it into end products that are safer. Phytoremediation, a low-cost, solar-energy-driven cleanup technique, is a similar concept, but involves using plants to pull contaminants from the soil and water.

These ideas offer hope for a better future where we value the environment more, damage the land less—and realize, finally, that the Earth itself is a limited and precious resource.

SOURCES

"Amazon Rainforest Vanishing at Twice Rate of Previous Estimates." *Guardian*, October 21, 2005, https://www.theguardian.com/environment/2005/oct/21/brazil.conservationand endangeredspecies.

Calero, Robee. "Land Pollution." Slideshare, https://www.slideshare.net/EeborSaveuc/land-pollution-49597029.

Chakrabarti, Vishaan. "This Land Is Our Land." *Urban Ominbus*, July 21, 2010, https://urbanomnibus.net/2010/07/this-land-is-our-land/.

Chen, Daniel. "Area of the Earth's Land Surface." *The Physics Factbook: An Encyclopedia of Scientific Essays* (2001), https://hypertextbook.com/facts/2001/DanielChen.shtml.

"Climate Change Threats and Solutions." The Nature Conservancy, https://www.nature.org/ourinitiatives/urgentissues/global-warming-climate-change/threats-solutions/index.htm.

Das, Surajit, and Hirak R. Dash. "Microbial Bioremediation: A Potential Tool for Restoration of Contaminated Areas." *Science Direct*, October 10, 2016, https://www.sciencedirect.com/science/article/pii/B9780128000212000017.

Davies, Wayne D. "Resilient Cities: Coping with Natural Hazards." In *Theme Cities: Solutions for Urban Problems*. GeoJournal Library, vol. 112. Dordrecht: Springer, 2015. Springer Link, https://link.springer.com/chapter/10.1007/978-94-017-9655-2_9.

"Deforestation and Its Extreme Effect on Global Warming." *Scientific American*, https://www.scientificamerican.com/article/deforestation-and-global-warming/.

"Deforestation: Compromises of a Growing World." Conserve Energy Future, https://www.conserve-energy-future.com/causes-effects-solutions-of-deforestation.php.

"Environmental Effects of Mining." Revolvy,https://www.revolvy.com/topic/Environmental%20effects%20of%20mining&item_type=topic.

Faith. "Conservation News: Forests, Snow Leopards, and Lions." *Animal Almanac* (blog), January 28, 2017, https://animalalmanacblog.wordpress.com/tag/half-of-the-worlds-rainforest-already-gone.

"Forest Facts." American Forests, http://www.americanforests.org/explore-forests/forest-facts/.

"Fossil Fuel Power Station." WikiVisually, https://wikivisually.com/wiki/Fossil_fuel_power_station.

Freudenrich, Craig. "How Landfills Work." *How Stuff Works*, https://science.howstuffworks.com/environmental/green-science/landfill6.htm.

"Hazardous Waste D+." Infrastructure Report Card, https://www.infrastructurereportcard.org/cat-item/hazardous-waste/.

Jindal, Sahil. "Report on Land Pollution." Slideshare, https://www.slideshare.net/funkyguysahil/report-on-land-pollution.

"Land Pollution Facts and Statistics." *Rubbish Please* (blog), January 28, 2016, https://rubbishplease.co.uk/blog/land-pollution-facts-statistics/.

"Learn about Landfills." Advanced Disposal, http://www.advanceddisposal.com/for-mother-earth/education-zone/learn-about-landfills.aspx.

"Mine Tailings." Ground Truth Trekking, October 10, 2014, http://www.groundtruthtrekking.org/Issues/MetalsMining/MineTailings.html.

"Native Plant, Invasive Plant . . . What Are the Differences? Why Does It Matter?" Ecological Landscape Alliance, https://www.ecolandscaping.org/native-and-invasive-plants/.

"Our World's Largest Rainforest: The Amazon." World Wildlife Fund (video), https://www.worldwildlife.org/videos/our-world-s-largest-rainforest-the-amazon#.

"Paris 2015: Tracking Country Climate Pledges." Carbon Brief, https://www.carbonbrief.org/paris-2015-tracking-country-climate-pledges.

"People in Antarctica." *Origins*, http://exploratorium.edu/origins/antarctica/people/index.html.

"Report on Land Pollution." Study Mode Research, 2013, http://www.studymode.com/course-notes/Report-On-Land-Pollution-1645259.html.

Roberts, Jon. "Garbage: The Black Sheep of the Family: A Brief History of Waste Regulation in the United States and Oklahoma." Oklahoma Department of Environmental Quality, 2015, http://www.deq.state.ok.us/lpdnew/wastehistory/wastehistory.htm.

Sherlock, Ruth. "Too Late: 400 Cities in America, Including Miami and New Orleans, Will Likely Be Submerged by Rising Sea Waters." *Telegraph*, October 15, 2015, http://www.businessinsider.com/cities-that-will-be-underwater-because-of-climate-change-2015-10.

"Soil Quality." Grace Communications Foundation, http://www.sustainabletable.org/207/soil-quality.

Tangahu, B. V. et al. "A Review of Heavy Metals (As, Pb, and Hg) Uptake by Plants through Phytoremediation." *Hindawi International Journal of Chemical Engineering* (August 2011), https://www.hindawi.com/journals/ijce/2011/939161/.

"Toxic Waste." Revolvy, https://www.revolvy.com/topic/Toxic%20wast.

"U.S. Fire Statistics." U.S. Fire Administration, https://www.usfa.fema.gov/data/statistics.

Vidal, John. "We Are Destroying Rainforests So Quickly They May Be Gone in 100 Years." *Guardian*, January 23, 2017, https://www.theguardian.com/global-development-professionals-network/2017/jan/23/destroying-rainforests-quickly-gone-100-years-deforestation.

Vitali, Ali. "Trump Pulls U.S. Out of Paris Climate Agreement." *NBC News*, June 1, 2017, https://www.nbcnews.com/politics/white-house/trump-pulls-u-s-out-paris-climate-agreement-n767066.

Vyhnak, Carola. "Is Sewage Fertilizer Safe?" *Toronto Star*, July 12, 2008, https://www.thestar.com/life/health_wellness/2008/07/12/is_sewage_fertilizer_safe.html.

"What Are Tailings? Their Nature and Production." Tailings.info, http://www.tailings.info/basics/tailings.htm.

"What Are the Main Causes of Soil Degradation?" Share Your Knowledge, http://www.biologydiscussion.com/soil/what-are-the-main-causes-of-soil-degradation/7276.

"What Is Food Pollution?" Environmental Pollution Centers, https://www.environmentalpollutioncenters.org/food/.

"What Is Habitat Destruction? Effects, Definition & Causes." Study.com, https://study.com/academy/lesson/what-is-habitat-destruction-effects-definition-causes.html.

"What Is Land Pollution." Conserve Energy Future, https://www.conserve-energy-future.com/causes-effects-solutions-of-land-pollution.php.

Wood, Kimberley. "The Mysterious Case of the Vanishing Rainforest (and How You Can See It Before It's Too Late)." *The Gist*, February 17, 2017, https://the-gist.org/2017/02/the-mysterious-case-of-the-vanishing-rainforest-and-how-you-can-see-it-before-its-too-late/.

Woodford, Chris. "Land Pollution." *Explain That Stuff*, March 27, 2018, http://www.explainthatstuff.com/land-pollution.html.

5

SPACE JUNK

> Thank God man cannot fly and lay waste to the sky as well as the Earth.—*Henry David Thoreau*

The good news is that no one has ever been injured or killed due to falling space junk. The bad news is that unless we clean up our extraterrestrial neighborhood, space travel, especially in orbit, is very likely going to be deadly. In space are thousands of collisions with space junk waiting to happen.

NASA claims that on average at least one piece of space debris has fallen back to Earth each day for the past fifty years, including parts of various space stations, satellites, pieces of rockets, and a variety of space waste from the size of a bus to a paper clip, including a Soviet nuclear reactor. The only person known to have been hit by space debris was Lottie Williams of Oklahoma, who was struck on the shoulder by part of a fuel tank from a Delta rocket—she said she hardly felt the impact.

In 1979, a kid from Australia got luckier. As Skylab, the first orbiting U.S. space station, was facing a rugged reentry, some American newspapers jokingly proposed "Skylab insurance," which would pay subscribers for death or injury due to careening fragments. The *San Francisco Examiner* specifically offered a $10,000 prize to the first person to deliver a piece of Skylab debris to its office within 72 hours of the crash. They figured that the satellite wasn't coming down anywhere near the continental United States, as NASA claimed, so the newspaper felt it was making a safe bet.

The reward caught the eye of 17-year-old Stan Thornton, of Esperance, Australia, who awoke to a commotion when Skylab's space scrap boomed its way into the atmosphere over his house, and pelted his property with fragments. He grabbed a few charred bits from his yard and hopped on a plane to the United States and the *Examiner*'s office as soon as he could to get there before the deadline. His pieces were examined by NASA, which declared they were genuine. So, not only was no one hurt, Stan was $10,000 richer.

Our immediate celestial neighborhood is getting more unkempt and congested day by day, just one more example of how we are not great caretakers of our environments. We really ought to be ashamed, yet we are doing nothing about it. It took us centuries to trash Earth, but only a few decades to litter the space around it. And it's more than about time to clean up our act. Or, as mom would say, "Pick up your space!"

As of mid-2016, there were about 780 satellites in the region of low Earth orbit, with more planned all the time, according to the Union of Concerned Scientists. Orbiting satellites share space with about 500,000 pieces of junk half an inch across and larger, according to NASA's estimates. Because everything is traveling at about 17,500 mph, something the size of a grain of rice can easily puncture solar arrays, and a piece that's an inch or two across can crash into a spacecraft with more energy than a bullet from a .50 caliber machine gun.

The higher the altitude, the longer the orbital debris will typically remain in Earth's trajectory. Fragments left in orbit below 380 miles (600 kilometers [km]) normally fall back to Earth within several years. At altitudes of about 500 miles (800 km), the time for orbital decay is often measured in decades. Above 621 miles (1,000 km), orbital debris will normally continue circling the Earth for a century or more.

There are different types of what can be called space junk. Natural objects like asteroids and comets are chunks of flying space rock left over from the formation of the solar system about 4.5 billion years ago. Some are as massive as mountains, most are smaller than pea gravel. If one big enough ever hits Earth, as Bill Nye the Science Guy says, "It's control-alt-delete for civilization." But we denizens of Earth aren't responsible for the wreckage of the big bang.

LUNAR LEFTOVERS AND A MARTIAN MESS

We are responsible for the mess of "orbital debris" that we left by accident, neglect, or design, and dumped on the Moon and Mars—our newest extraterrestrial spacefills. Humans have a real nasty habit of discarding their stuff wherever they go. According to NASA, hundreds of millions of pieces of space debris are now floating through our region of the solar system, and there are currently more than 400,000 pounds of junk littering the moon.

It's not all just moon junk, though, some devices are still functioning, and some are there for sentimentality's sake. One is the silicon plaque left by the *Eagle* astronauts in the Sea of Tranquility bearing goodwill messages from U.S. presidents Eisenhower, Kennedy, Johnson, and Nixon and the leaders of 73 nations, along with—of course—some names of members of Congress. It reads, "Here men from the planet Earth first set foot upon the Moon. July 1969, A.D. We came in peace for all mankind." And then we left our garbage.

Most of that discarded stuff lies off to the west of the actual landing site, in an arc of space-litter known as the "Toss Zone." The junkyard on the moon includes 73 moon probes and unmanned space vehicles. The three lunar roving vehicles from Apollo 15, 16, and 17 also sit on the surface of the moon. In addition, there are a couple of gold-colored balls along with five American flags. An aluminum sculpture memorial for fallen astronauts and cosmonauts is there as well, along with a mini art gallery. The works by various artists were displayed on a tiny penny-sized ceramic chip that included the famous doodle Andy Warhol drew of a penis.

Apollo 14's Edgar Mitchell returned with one of 12 Hassleblad cameras, which he later tried to sell at auction, but selfish NASA sued and recovered it. A gold-plated telescope designed by Dr. George Carruthers was left—the only telescope to make observations from the surface of another planetary body.

Elon Musk is the newest name in contributing to the junkyard of space. He may be a way-cool space entrepreneur, but he's debuting in space as a high-class litterer, contributing to the accumulation of space scrap in the next phase in Earth's interplanetary journeys. In 2018, SpaceX, Musk's company, launched a cherry-red Tesla Roadster and its dummy "driver," Star Man, on its Falcon Heavy rocket, blasting David

Bowie's "Life on Mars" on its way to Mars's orbit. We don't mean to be a spoil-space-sport about this, but come on—the car isn't exactly going to be "draggin' the cosmos" and doing hot laps on the red planet. The closest it's going to get is about 4.3 million miles away.

Next dump—Mars. Most of the debris there is spacecraft and smaller parts, such as springs, parachutes, and heat shields. As of November 5, 2016, there had been about 14 spacecraft missions to the surface of Mars, some of which left spacecraft scrap. One can only guess what's to come with colonization.

The biggest offender of space (and Earth) litter is China, a latecomer to space, but accounting for more than 31 percent of all space junk. In one fell swoop, they shot down one of their dying satellites, leaving behind thousands of objects that are in low Earth orbit, waiting for fragmentation impact.

This is not only a real threat to space stations, but also any astronauts, cosmonauts, or other spacenauts working up there. Dire consequences are forecast if we don't get our poop in a pile and cleaned up. Donald Kessler, who started the Environmental Effects Project for the NASA Johnson Space Center, says that near outer space has become Earth's largest off-terrestrial dump site.

The moon, our next-door neighbor, has felt the negligent hand of humans with our various detritus, including astronaut poop and barf bags.

This, of course, has conservationists very concerned about the moon's protection should further trips there be undertaken. At present, Tranquility Base is still tranquil because there is no wind or rain up there to damage or blow things around or, at the moment, anymore tourists to leave things behind. But it looks as if not much thought has been given to protection of our rocky satellite or laws enacted regarding its protection. As yet, only California and New Mexico have recognized the moon as an international historic sight, and for now, it isn't recognized as a Historical Landmark or considered a World Heritage Site by the United Nations Educational, Scientific and Cultural Organization (UNESCO).

Under international (interplanetary?) law, the U.S. government still owns everything it left on the moon. And apparently NASA is not too sentimental about souvenirs, unless an astronaut brings one home. When their projects end up there they tend to get a little loose about aftereffects. And in the event of cleanup and/or damages by our alien space litter, who gets stuck with the tab? If space debris causes damage to persons or property in other countries, the Convention on International Liability for Damage Caused by Space Objects obligates the U.S. government, if it's ours, to pay for it. And believe it or not, most homeowner insurance protects against damage caused by falling objects from space—which might also be considered valuable collector's booty to boot.

Interest in junkets to the moon has perked up again. Russia, India, and China plan to send robotic landing craft and other space treks are undoubtedly being planned by private tour guides. Space impresario Elon Musk, among others, hopes to send 4,000 satellites into low Earth orbit—where most of the space junk is located—to provide internet service worldwide, and also offer space jaunts to the moon—or at least junkets in low orbit. If those cosmos tour guides get their wish, amateur astronauts in rented moon rovers will be rolling over Neil Armstrong's footprints soon, and souvenir stands will sprout up at Tranquility Base.

ASSESSING THE PROBLEM

Don Kessler originally wanted to become an astronaut, but his attention shifted from space launches to space waste. He wondered about collisions in space and what will happen to the possible multiple ruins that we leave in space. He answered the question in a landmark paper in 1978, titled "Collision Frequency of Artificial Satellites," detailing the science behind what is now unofficially known as the Kessler Syndrome. He postulated that space junk collides with other space junk, producing more and more fragments, until the debris eventually renders low Earth orbit impassable. He also scolded us for being space litterbugs and leaving behind all manner of refuse from large derelict spacecraft to tiny paint chips (NASA frequently had to replace Space Shuttle windows damaged by orbiting flakes of paint).

His predictions of heavenly heavyweight fender-benders made him big-time points in the space debris field and a space junk celebrity and authority. After the paper was published, it made NASA and others ponder the effects of their collective crap circling the planet, and they created the Orbital Debris Program Office (ODPO) to study the problem and see what they could do about it.

With the usual alacrity of most governmental agencies, they chose to wait around expressing the fact that this was probably going to be expensive and it was too far in the future to get freaked out about. Then, in February 2009, two satellites clipping along at five miles per second smacked into one another, smashing each other into smithereens, the first time a "head-on" in space was recorded and studied. Each of the 2,100 pieces of the leftover crash is capable of causing another spaceship wreck that could menace some high-priced hardware, not to mention lives. Debris shields on spacecraft are only effective in withstanding impacts of particles smaller than half an inch.

Space scrutiny is a crucial part of the U.S. Space Command (USSPACECOM) mission and involves detecting, tracking, and cataloging human-made objects orbiting Earth, such as active and inactive satellites, spent rocket bodies, and fragmentation debris. NASA and the Department of Defense (DOD) cooperate and share responsibilities for monitoring orbital debris. Using special ground-based sensors and inspections of returned satellite surfaces, NASA statistically determines the extent of the population up there for objects less than four inches (10 cm) in diameter, and the DOD's Space Surveillance Network, run by the Joint Operations Center at Vandenberg Air Force Base, tracks discrete objects as small as two inches (5 cm) in diameter in low Earth orbit and about 40 inches in geosynchronous orbit (synchronous with the Earth's rotation). The Network has some important tasks:

- Predict when and where a decaying space object will reenter Earth's atmosphere.
- Prevent a returning space object, which to radar looks like a missile, from triggering a false alarm and missile attack, preempting WWIII.
- Chart the present position of space objects and plot their anticipated orbital paths.
- Detect new human-made objects in space.
- Determine which country owns a reentering space object.
- Inform NASA whether or not objects may interfere with the Russian International Space Station orbit.

There are several more agencies and programs keeping track of what's loose and possibly ready to rain back to Earth, or cause another "accident." The Space Surveillance Network tracked an average of five possible collisions or "conjunctions" per day as of 2009. A year later, they began tracking 1,000 satellites and rocket pieces and another 15,000 items the size of a Mini Cooper, looking for a possible 75 pileups per day. There are more than half million smaller pieces that they can't track because of size but that are still capable of damaging or knocking down a satellite.

So far, all we have had to deal with are close calls, like in 2001 when the $100 billion Russian International Space Station got a heads-up warning to prepare for possible impact, making the cosmo/astronauts get aboard a "life raft" capsule. It missed by a few miles, a very close call in space. Even a relatively small piece of detritus could have the impact of almost 15.5 pounds of TNT—far more than needed to "total" a space station.

Needless to say, if a spacecraft were hit with a softball-size scrap of metal traveling at 17,500 miles per hour, it could do some critical damage. For example, in 2009, a retired Russian satellite collided with a U.S. commercial satellite, and the results were catastrophic. The collision destroyed both satellites, adding more than 2,000 pieces of space scrap to the growing mass.

To stop this collision-cascading process, we have to be able to capture and return or safely dispose of satellites, and we don't have the ability to do that. If we can't figure out how to return large satellites to Earth, then, Kessler says, we'll just have to start picking up all the pieces.

The amount of our space clutter is expected to double, even triple over the next few years as "end-of-life" objects multiply and old ship wrecks in space escalate the possibility of more collisions, like so many pool balls banging around in the heavens. One senior NASA scientist says ominously, "We're in a runaway environment and we won't be able to use space in the future if we don't start dealing with this now."

However, so far NASA has been somewhat cavalier about the situation and refers the problem as "The Big Sky Theory" figuring that space is so vast it is fine to just leave stuff and wait for it to descend into a degraded slope simply letting our atmosphere vaporize the stuff.

In 1979, Arthur C. Clark, the noted scientist and author, wrote in a rather ominous tone that the space accumulation, "has to be located and somehow disposed of," because if not, "the Earth would be cut off losing the ability to communicate and we would sink back into a dark age, resulting in chaos, disease, starvation, and destroying much of the human race."

Well, even though this statement sounds a little like heavy-handed doomsday-prediction fiction, Kessler and some of his associates at NASA are taking it seriously. He predicts that eventually, there will be so much space junk that leaving Earth to explore deep space will be impossible. That includes sending satellites to distant stretches of the solar system and manned missions to Mars.

Kessler and NASA have developed guidelines slowing the buildup of space debris, suggesting that there should be rules of celestial right-of-way, about what and how much could be abandoned, who would be responsible, and whether higher "graveyard" orbits would supposedly be out of harm's way.

CLEANING UP OUR SPACE

In 2009, NASA, and the Defense Advanced Research Projects Agency (DARPA), an arm of the U.S. DOD responsible for the development of emerging technologies, convened a space junk wake-up call. The team took a hard look at the issues and challenges of decluttering space of human-made orbital rubble. More than 50 presentations from the United States, Russia, France, Germany, and Japan addressed not only the technical and economic challenges, but also the legal and policy issues associated with orbital debris removal.

Some of the suggested plans sound kind of spacey:

- Launching sticky balls that objects would adhere to and then dragging them out of orbit, gently floating them into large nets and then into the Earth's atmosphere where they would vaporize.
- A giant inflatable donut that would bounce space junk into the atmosphere below.
- Small nanosatellites that would gather up the detritus.

- Electrifying space rubbish and attracting it to the Earth's magnetic field, where it would burn up.
- Huge "aerogel-laden" sticky puffballs to snare debris.
- Various types of galloping gotcha-tethers.
- Vacuum cleaner–type contraptions.
- A scheme for rapid-fire laser pulses to zap off a micro-thin surface layer of targeted debris that acts as a miniature rocket motor—with enough "oomph" to tease the object's orbit and reentering it in a fiery finale.

Another idea is called Round Up Space Trash Low Earth Orbit Remediation (RUSTLER), which makes use of an electrodynamic tether along with two other unconventional technologies to enable safe and cost-effective removal of space junk. Still another was proposed in 2014 and is still in development by the European Space Agency (ESA). It involves two possible mechanisms of capture: nets and/or robotic arms.

One more concept aims to pick off space debris one by one. The CleanSpace One project is being developed to de-orbit Switzerland's SwissCube Nanosat. The satellite cleanup device is set to launch from the SOAR space plane, an unmanned mini-shuttle. The CleanSpace One also looks to fling the target satellite into the atmosphere to incinerate it. Whichever is chosen, the goal will be the same: to snag the debris out of orbit and bring it to a lower altitude into the Earth's atmosphere, where it will burn up. Finally, and perhaps the most fun, is sending balloons up to space to hit debris with a gust of compressed solar wind sending it hurtling toward the atmosphere. An additional proposal is to employ a giant plastic-wrap apparatus that will enfold space debris and solar-sail it into Earth's atmosphere.

In 2017, Japan's space agency, JAXA, launched an electrodynamic tether, called an EDT, into space. The cable is 700 meters (2,296 feet) long and meant to assist in de-orbiting dangerous space junk. The idea is that the electrified tether will work with the attached 44-pound counterweight to slow down debris and redirect it toward Earth's atmosphere, where it will safely be destroyed.

In the future, in the order to pick up the tab for the cleanup there will have to be an assessment, like a space tax, on commercial satellite companies to offset costs. NASA received $5 billion to attack the problem, but it's going to take a lot more. Eventually satellites will be required to be equipped with built-in disposal devises. It's been suggested that the surest path to resolving the problem is to establish a system that places a high price on anyone generating new debris, while creating rewards for anyone who lessens it.

At present, it seems that you can't remove or even move somebody else's space flotsam and jetsam without the owner's permission. Anyone or anything grabbing a piece of debris and bringing it down could be considered poaching, especially if secret or proprietary technology is involved.

But whatever the method, it's time we started to clean up our act. Maybe that's why we haven't had direct contact with any extraterrestrial beings—they just can't stand our mess.

SOURCES

"2009 Satellite Collision." Revolvy, https://www.revolvy.com/topic/2009%20satellite%20collision&item_type=topic.

Betz, Eric. "Science Guy Bill Nye Makes Asteroids Funny." *Arizona Daily Sun*, April 18, 2013, http://azdailysun.com/news/local/science-guy-bill-nye-makes-asteroids-funny/article_5a8ffa67-2004-5c65-b501-82964af11cb6.html\.

Chavers, Marcus. "A Chilling Visualization of Space Debris and the Ideas to Remove It." *News Ledge*, January 8, 2017, https://www.newsledge.com/space-debris/.

David, Leonard. "NASA, DARPA Host Space Junk Wake-Up Call." *Fox News*, December 9, 2009, http://www.foxnews.com/tech/2009/12/09/nasa-darpa-host-space-junk-wake.html.

David, Leonard. "Space Junk Menace: How to Deal with Orbital Debris." *Space.com News*, January 25, 2013, https://www.space.com/19445-space-junk-threat-orbital-debris-cleanup.html.

"Donald J. Kessler." Wikipedia, https://en.wikipedia.org/wiki/Donald_J._Kessler.

Elkins-Tanton, Lindy. "Space Exploration Isn't Just about Science." *Slate*, March 7, 2017, http://www.slate.com/articles/technology/future_tense/2017/03/if_someone_beats_the_u_s_to_mars_it_will_feel_like_a_military_defeat.html.

Emspak, Jesse. "How Can Humans Clean Up Our Space Junk?" *The Verge*, December 30, 2016, https://www.theverge.com/2016/12/30/14116918/space-junk-debris-cleanup-missions-esa-astroscale-removedebris.

Griggs, Mary Beth. "Tiny Debris Chipped a Window on the Space Station." *Popular Science*, May 12, 2016, https://www.popsci.com/paint-chip-likely-caused-window-damage-on-space-station.

Gupta, Vishakha. "Critique of the International Law on Protection of the Outer Space Environment." *International Journal of Space Politics & Policy* 14, no. 1 (2016): 20–43.

Hanes, Elizabeth, "The Day Skylab Crashed to Earth: Facts about the First U.S. Space Station's Re-Entry" *History*, July 11, 2012, https://www.history.com/news/the-day-skylab-crashed-to-earth-facts-about-the-first-u-s-space-stations-re-entry.

"History of the Earth." Wikipedia, https://en.wikipedia.org/wiki/History_of_Earth.

"How to Clean Space: Disposal and Active Debris Removal." *Aerospace*, December 10, 2015, http://www.aerospace.org/crosslinkmag/fall-2015/how-to-clean-space-disposal-and-active-debris-removal/.

Howell, Elizabeth. "Space Junk Clean Up: 7 Wild Ways to Destroy Orbital Debris." *Space.com*, March 3, 2014, https://www.space.com/24895-space-junk-wild-clean-up-concepts.html.

"How Long Will Orbital Debris Remain in Earth Orbit?" NASA, 2011, https://www.nasa.gov/news/debris_faq.html.

"Humans Have Already Left More Than 400,000 Pounds of Junk on the Moon." *Your Daily Science Source*, 2018, http://sciencetrend.org/space/humans-have-already-left-more-than-400000-pounds-of-junk-on-the-moon/.

"Joint Space Operations Center." Wikipedia, March 29, 2018, https://en.wikipedia.org/wiki/Joint_Space_Operations_Center.

Kaye, Ben. "Elon Musk's Latest SpaceX Rocket Is Sound Tracked By an Infinite Loop of David Bowie's 'Space Oddity.'" *COS*, February 6, 2018, https://consequenceofsound.net/2018/02/elon-musks-latest-spacex-rocket-is-soundtracked-by-an-infinite-loop-of-david-bowies-space-oddity/.

Kushner, David. "Five Ideas to Fight Space Junk." *Popular Science*, July 13, 2010, https://www.popsci.com/technology/article/2010-07/cluttered-space.

La Von, Michelle. "Kessler Syndrome." *Space Safety Magazine*, 2014, http://www.spacesafetymagazine.com/space-debris/kessler-syndrome/.

"List of Artificial Objects on Mars." Wikipedia, November 2016, https://en.wikipedia.org/wiki/List_of_artificial_objects_on_Mars.

"List of Artificial Objects on the Moon." Wikipedia, 2015, https://en.wikipedia.org/wiki/List_of_artificial_objects_on_the_Moon.

Long, Tony. "Heads Up, Lottie! It's Space Junk!" *Wired*, January 22, 1997, https://www.wired.com/2009/01/jan-22-1997-heads-up-lottie-its-space-junk/.

McMahon, Ed. "The Moon Is Cluttered with Our Stuff." *Jewel* 92, July 24, 2017, http://jewel92.com/moon-cluttered-stuff/.

NASA Wiki. http://nasa.wikia.com/wiki/Apollo_11.

"NASA's Recommendations to Space-Faring Entities." NASA, July 20, 2011, https://www.nasa.gov/pdf/617743main_NASA-USG_LUNAR_HISTORIC_SITES_RevA-508.pdf.

Redd, Nola Taylor. "Space Junk: Tracking & Removing Orbital Debris." *Space.com*, March 8, 2013, https://www.space.com/16518-space-junk.html.

Schwartz, Evan I. "The Looming Space Junk Crisis: It's Time to Take Out the Trash." *Wired*, May 24, 2010, https://www.wired.com/2010/05/ff_space_junk/.

Shapiro, Jeffrey Scott. "Russian Space Junk Sends Space Station Astronauts into 'Safe Haven' Vehicle." *Washington Times*, July 16, 2015, https://www.washingtontimes.com/news/2015/jul/16/space-station-astronauts-dodge-russian-space-junk/.

Smith, Kiona N. "The Apollo 11 Astronauts Left a Lot of Junk on the Moon." *Forbes*, July 20, 2017, https://www.forbes.com/sites/kionasmith/2017/07/20/the-apollo-11-astronauts-left-a-lot-of-junk-on-the-moon/#4c7bfb324ca0.

"Space Debris." Wikipedia, https://en.wikipedia.org/wiki/Space_debris.

http://blogs.discovermagazine.com/80beats/2009/03/12/space-junk-near-miss-sends-astronauts-scrambling-into-escape-pod/#.Wtp7yTNlDcs.

"Space Property: Who Owns It?" *Science Focus*, August 1, 2011, http://www.sciencefocus.com/feature/health/who-owns-space.

"Space Situational Awareness." European Space Agency, November 11, 2017, http://www.esa.int/Our_Activities/Operations/Space_Situational_Awareness/Space_Surveillance_and_Tracking_-_SST_Segment.

"Space Surveillance." United States Space Command, http://www.au.af.mil/au/awc/awcgate/usspc-fs/space.htm.

Stansbery, Gene. "NASA's Orbital Debris Program Office." http://www.au.af.mil/au/awc/awcgate/usspc-fs/space.htm.

Stansbery, Gene, and Nicholas Johnson. "Orbital Debris: Past, Present, and Future." NASA Orbital Debris Program, https://ntrs.nasa.gov/archive/nasa/casi.ntrs.nasa.gov/20130000305.pdf.

Strickland, Eliza. "Space Junk Near-Miss Sends Astronauts Scrambling into Escape Pod." *Discover*, March 12, 2009.

Vallancey, Sam. "Elon Musk Plans to Launch 4000 Satellites." *Indy Tech* September 14, 2015, https://www.independent.co.uk/life-style/gadgets-and-tech/news/elon-musk-plans-launch-of-4000-satellites-to-bring-wi-fi-to-most-remote-locations-on-earth-10499886.html.

"Wait, What?" DARPA, December 9–11, 2015, http://archive.darpa.mil/WaitWhat/.

Wenz, John. "What a $19 Billion Budget Will Buy NASA." *Popular Mechanics*, February 9, 2016, https://www.popularmechanics.com/space/news/a19368/what-18-billion-will-get-a-space-agency-like-nasa/.

Whitwell, Laurie. "NASA Forces Apollo Astronaut to Give Back Space Camera He Had Brought Home from 1971 Moonwalk Mission as Souvenir." *DailyMail*, October 2011, http://www.dailymail.co.uk/news/article-2055433/NASA-forces-Apollo-astronaut-Edgar-Mitchell-space-camera-brought-home-1971.html.

Williams, Matt. "China Has a Plan to Clean Up Space Junk with Lasers." *Universe Today*, January 16, 2018, https://www.universetoday.com/138263/china-plan-clean-space-junk-lasers/.

6

FISH TO FARM TO TABLE TO TRASH

> Imagine walking out of a grocery store with four bags of groceries, dropping one in the parking lot, and just not bothering to pick it up. That's essentially what we're doing.
>
> —*Dana Gunders*

People don't know whether to be astonished, ashamed, or amused when it comes to the magnitude of food that Americans throw away. No other country in the history of the world has the ability to raise, produce, eat, and toss out as much as we do.

Mothers still shake fingers at their children admonishing fussy eaters, "Just think of all the children in the world that are starving and would love this food you're wasting." Of course, most of us answered, "OK, send it to them." Fact is, mom was right. And so were we, kind of, because we export more food than any other nation on Earth, more than $135 billion each year.

And all levels of the food system are riddled with waste—farming, harvesting, transportation, packaging, wholesale and retail marketing, and finally our tables. Food waste levels are 20 to 25 percent of manufacturing, 15 to 20 percent of retail sales, and 55 to 65 percent from consumers.

The truth is there is enough food to feed every single person in America if we can help farmers, manufacturers, and retailers get the food to the people who need it. That's why the Feeding America network and its partners work with farmers and food companies to rescue food and deliver it to families facing hunger. Things could be changing on the food waste front, with the U.S. Department of Agriculture (USDA) and the Environmental Protection Agency (EPA) setting the first-ever national food waste reduction goal, aimed at cutting national food waste in half by 2030.

Food facts and figures:

- Less than 1 percent of pesticides applied to crops reach the pest; the rest poisons the ecosystem. Each year, 25 million people are poisoned by pesticides in less-developed countries, and more than 20,000 die.
- One-third of the world's fish catch and more than one-third of the world's total grain output are fed to livestock—the most wasteful way to produce protein. It takes 175 gallons of water to produce an eight-ounce soy burger, and a whopping 450 gallons to produce a quarter-pound beef burger.
- About 50 percent more food is wasted today per person than in the mid-1970s and more than two-thirds of the food we throw away is edible.
- Food is so cheap and available to most families that they throw out up to 25 percent of their food and beverages. This can cost the average family between $1,365 and $2,275 annually.
- A study by the University of Arizona found that 14 percent of the food trashed in America was not even opened.
- A 20 percent reduction in food waste would be enough to feed 25 million Americans. Five percent of Americans' leftovers could feed four million people for one day.
- Between 40 and 50 percent of wasted food uses up 25 percent of all fresh water in the United States.
- It costs $750 million to dispose of food thrown away annually.
- Dumpster divers are part of the "Freeganism" movement. Many say that they would never be able to afford such good-quality food without pilfering "garbage."
- America's convenience stores, restaurants, and supermarkets throw out more than 27 million tons of edible food worth $30 billion a year.
- Approximately 12 percent of the food served as part of the National School Lunch Program is wasted, resulting in an estimated economic loss of $600 million per year.

- By 2050, Americans will be consuming 3,070 calories a day, yet the average man requires only 2,000 calories, or less, women need about 1,800, and small children about 1,000. Cut out 500 calories a day and you could lose 25 pounds over a year's time.
- According to the USDA, the top three food groups wasted in 2010 were dairy products (25 billion pounds, or 19 percent); vegetables (25 billion pounds, or 19 percent); and grain products (18.5 billion pounds, or 14 percent), at a cost of $161.6 billion.

WASTE MORE, WANT MORE

A new study from the University of Arizona in Tucson says one of the biggest wastes of energy in America could be the food in your garbage can. Even the most sustainably farmed food does no good if the food is never eaten. The bottom line of our garbage says that 40 to 50 percent of all food ready for harvest never gets eaten—from farmers, supermarkets, restaurants to your own kitchen.

Each year, American food waste represents the energy equivalent of 350 million barrels of oil, according to new research from the University of Texas. Just getting food from farm to folks eats up to 15.7 percent of the total U.S. energy budget, uses 50 percent of our farmland, and swallows close to 80 percent of all freshwater consumed in the country. Reducing food losses by just 15 percent would be enough to feed one in six Americans that lack a secure supply of food, according to the U.S. Energy Information Administration (EIA).

This mountain range of wasted food totals about 63 million tons, of which more than 10 million tons never get harvested from farms and 52.4 million tons ends up uneaten, according to the End Hunger Network.

Every ton of food wasted results in four times its amount in greenhouse gas emissions and reducing food waste by half would prevent as much climate-affecting gases as taking every other car off the road. The University of Arizona believes that if Americans cut their food waste in half, it would reduce the country's environmental impact by 25 percent. Think about that fact every time you plan to supersize at your favorite restaurant, or casually chuck out yesterday's leftovers.

Noncomely Comestibles—a Corrupted Cornucopia

Produce in America demands cover-girl-looking produce that causes farmers to leave food unharvested in the field and retailers and wholesalers to throw out perfectly edible food. Americans are squandering the equivalent of $165 billion each year by turning up their collective noses at what gets turned back into the earth by farmers as very expensive fertilizer.

A report from the Natural Resources Defense Council (NRDC) that tracks food waste from "farm to fork," states that from 7 to 50 percent of the produce that's grown in the country simply gets stranded on fields each year. According to the charity Feeding America, more than 6 billion pounds of fruits and vegetables go unharvested or unsold each year, much because of cosmetic quirks.

Once crops have been harvested, culling is the primary reasons for losses of fresh produce. Culling is the removal of produce due to size, color, weight, blemish level, and Brix (a measure of sugar content). One large cucumber farmer estimated that fewer than half the vegetables he grows actually leave his farm and that 75 percent of those that remain are edible.

A large tomato-packing house reported that in mid-season it can fill a dump truck with 22,000 pounds of discarded tomatoes every 40 minutes. And a packer of citrus, stone fruit, and grapes estimated that 20 to 50 percent of the produce he handles is unmarketable but perfectly tasty. Although some of this "not-quality-grade" produce goes to processing plants, much does not.

A comprehensive 2012 report by the NRDC found that 43 billion pounds of food was thrown away just on the retail level alone. The USDA estimates that supermarkets toss out $15 billion worth of unsold fruits and vegetables each year.

According to a study by ReFED (some of the nation's leading business, nonprofit, foundation, and government leaders committed to reduce U.S. food waste), 20 billion pounds of fresh fruits and vegetables, and occasionally upward of 50 percent, for a particular crop, are discarded or left unharvested every year. Even processed food that doesn't pass muster is discarded, like peanut butter that is too chunky to be creamy, and too creamy to be chunky. Believe it or not, there are laws that make producers throw out potatoes because they are too small, cucumbers too curvy, and tomatoes that don't blush the right shade of red.

At farmer's markets, which have more than doubled in the past ten years, growers are allowed to sell good-quality products that might not meet size, specific shelf life, or cosmetic criteria imposed by retailers.

Some good news is that California recently passed a bill allowing growers to receive a tax credit for donations of excess produce to state food banks, joining Arizona, Oregon, and Colorado. California's America Gives More Act would allow small businesses, farmers, and restaurants to donate usable excess food to nonprofits that serve people in need. It would give them the ability to write off their donations, a boon that would enable Feeding America, the country's largest food donation organization, the ability to serve an additional 100 million meals a year to 49 million food-insecure people.

Date Waste

Then there's the issue of expiration dates, which can be wildly confusing. Food safety is rarely, if ever, impacted by those dates. There are more than 10 different variations of expiration dates; this confuses people and causes them to play it overly safe and toss out food too soon.

According to a study by Harvard University and the NRDC, one industry estimates that each supermarket throws out an average of $2,300 worth of food each day—milk, precut veggies, even Twinkies—which we thought would survive a nuclear war—because of sell-by, use-by, and expiration dates. In fact, many stores pull items two to three days before the sell-by date out of concern that their image of having fresh products will decay. And more than 80 percent of Americans throw out food, believing that those dates are related to food safety.

There's no federal law covering sell-by food dates, and just about every state has different rules on what info needs to be on the food labels. Some supermarket chains have taken it upon themselves to do their own sell-by markings. The bottom line is that unless and until clear guidelines are set for expiration labeling, use common sense and don't freak out and toss edible goods just because the sticker says so.

Two leading food industry associations, the Food Marketing Institute (FMI) and Grocery Manufacturers Association (GMA) have jointly released voluntary guidance to help standardize date labels and reduce confusion. This is an enormously helpful step in addressing the vast amount of food going to waste in stores and homes across the country. Walmart made a similar change, going from 47 sell by or use by date phrases down to one.

The FMI/GMA guidelines closely follow the recommendations from NRDC and the Harvard Food Law and Policy Clinic's landmark 2013 report, *The Dating Game: How Confusing Labels Land Billions of Pounds of Food in the Trash*. It establishes two standard phrases to be used with date labels, each with an associated definition: "Best if used by" to indicate peak quality and "Use by" when the date does indicate some level of increased risk after some time. It phases out the use of "sell by" dates that are visible to consumers, and allows for "freeze by" to be added if that information can help clarify how to handle the product.

Fast and Loose with Food

The USDA estimates households and food service operations (restaurants, cafeterias, fast food, and caterers) together wasted 86 billion pounds of food in 2008, or 19 percent of the total U.S. retail-level food supply. According to a study at the University of Arizona, food waste amounts to 10 percent in the fast food industry and about 3 percent in full-service restaurants. The same study estimated that the total food loss in the United States per day was close to 50,000 pounds in all full-service restaurants and 85,000 pounds in fast-food restaurants. Take French fries for example. Even though they are consistently prepared throughout the day, each batch sits for an average of less than ten minutes before being tossed. And for health reasons, much of it cannot be donated.

Another restaurant issue is the sheer size of the offerings. It's not only bodacious burgers and bottomless fries that have grown bigger in recent years, but also the plates and glasses with which the food is served. From 1982 to 2002, the average pizza slice grew 70 percent in calories, a chicken Caesar salad doubled, and chocolate chip cookies quadrupled. Today, portion size can be two to eight times larger than USDA or FDA recommended standard serving sizes. On average, diners leave 17 percent of meals uneaten, and 55 percent of these potential leftovers are not taken home and are tossed in dumpsters, even though we were all admonished to be "clean plate rangers" by our parents.

Cafeterias and buffets leave large amounts of food sitting out at all times, thus decreasing the eat-by time and creating shocking amounts of edible waste. According to Aramark, a dining company that serves about 500 schools in the United States, students waste 25 to 30 percent less food when they don't have the option of carrying it on a tray. When Iowa State University removed trays, the school saw a 10-percent reduction in wasted food.

In restaurant kitchens, food is lost due to being incorrectly prepared or spoiled. In addition, some loss is trim waste and some simply due to overproduction. Approximately 4 to 10 percent of food purchased by restaurants becomes kitchen loss, both edible and inedible, before reaching diners. Restaurants and large commercial food operations are clueless about how much and what kind of food they are throwing away—even when it could save them money. The cheaper the food, the less concerned people are about wasting it.

HELP FROM UNCLE SAM AND FRIENDS

The U.S. government should conduct a comprehensive study for food losses in our food system and establish national goals for waste reduction. State and local governments should help by setting targets and implementing food waste prevention campaigns. Businesses should start by understanding the extent and opportunity of their own waste streams, creating and adopting best practices.

Grocery Outlet, a $960 million business with 148 stores, has made a business out of selling closeouts, overruns, and less pretty produce, with 75 percent of their products coming from those streams. The Farm to Family program in California recovers more than 120 million pounds of cosmetically challenged produce per year from farms and packers for distribution to food banks. Stop and Shop, a grocery store chain, was able to save an estimated $100 million annually after an analysis of freshness, shrink, and customer satisfaction in their perishables department. And the dollar stores often offer produce that would otherwise go to waste.

The Farm to Family program of the California Association of Food Banks has pioneered an approach it calls concurrent picking, whereby workers harvest unmarketable produce. The program covers the costs of additional labor, handling, packaging, and refrigeration and receives fresh produce at a greatly reduced rate (10 to 15 cents per pound and more) and growers are able to deduct it as a charitable donation. This model also has the potential to serve secondary markets such as discount stores, afterschool snack programs, or other low-budget outlets and programs.

Biogas By-Product

Ironically, one of the solutions to dealing with food waste actually results in a product that improves air quality. Biogas is a type of biofuel that is naturally produced from the decomposition of organic waste and is an environmentally friendly energy source that can be cheaper than gasoline. Biogas (natural gas) vehicles produce less smog-related and greenhouse gas emissions compared to conventional vehicles. People who have switched their vehicles to biogas have saved around 40 to 60 percent compared to the price of gasoline.

Kroger, a supermarket chain, has undertaken a waste-to-energy process with a "food digester." It takes in surplus food and puts out biogas, providing 13 million kilowatt hours of electricity, or enough to power 2,000 homes for a year while using 150 tons of food a day that would normally be hauled to landfills.

IT'S UP TO YOU

- First of all, simply push away your plate and learn to eat less; obesity is rampant. Use smaller plates and don't pile your dish like it's your last meal. Use your leftovers.
- Encourage your local restaurant to sell half-portions and to offer smaller portions for the same price.
- Evaluate how to reduce waste before it's created, like planning meals in advance and buying only what is necessary at the supermarket.
- Save cooking fats and oils for alternative biofuels if there is a rendering plant close by.

- Compost discarded foodstuffs into a useful soil additive.
- An erasable board on the wall or the fridge will remind you of when you bought perishables, when you used them, and when to restock the fridge and the pantry.
- Freeze any items that won't be used soon.
- Funny-looking produce is perfectly good to eat.
- Try donating food that you are not going to eat. But be advised that many food banks do not take many kinds of donations for various reasons.
- Check out smart refrigerators that help monitor food use and storage.

Check out these apps:

- The Green Egg Shopper. This application helps you track your food purchases by how soon they will spoil.
- Love Food, Hate Waste. This app allows you to easily keep track of food planning, shopping, cooking meals, and making the most of leftovers.
- 222 Million Tons. This application helps you shop smarter.

Put the Waste to Work

According to a survey conducted by the National Waste & Recycling Association, 28 percent of Americans compost their food scraps, and this number is growing. Of those surveyed who do not compost, 67 percent said they would be open to trying it.

San Francisco became the first city in the United States to mandate composting with a goal of virtually zero landfill waste by 2020. Seattle followed suit, collecting 90,000 tons in the first year of its program and Austin is gearing up to institute mandatory composting for food businesses. Former New York mayor Michael Bloomberg proposed mandating composting in the city, and city planners are hoping it will lessen landfills by 30 percent. Many small towns like Mill Valley, California, offer free compost, created by their compostable garbage, to their customers.

NOT OFF THE HOOK: THE FISH STORY

Of the 600 marine fish stocks monitored by Fisheries and Aquaculture Organization (FAO):

- 47 percent of all edible seafood in the country goes to waste each year, according to a study by the Johns Hopkins Center for a Livable Future (CLF). Conservative estimates suggest the 2.3 billion pounds of seafood are squandered each year, enough to provide protein for more than 10 to 12 million people for an entire year.
- 40 to 60 percent of all fish are tossed as waste because of the wrong size, incorrect species, or some other reasons.

Fish are the last wild animal that people hunt in large numbers. Fish stocks are in infamously poor health and have been exploited, depleted, and polluted by waste in the last several decades to the extent that without urgent measures we may be the last generation to catch a reasonable amount of food from the oceans. Only 10 percent of the top predator species, such as tuna, shark, and swordfish, are left since industrial fishing began only 70 years ago. We are the only species that can eat or pollute another species out of existence.

Piscatorial Plastics

Fish caught off the coasts of California and Indonesia and sold in local markets have been found to have plastics and textile fibers in their guts, raising concerns over food safety. Plastic-eating fish are now showing up in supermarkets. Scientists at Ghent University in Belgium recently calculated that shellfish lovers are eating up to 11,000 plastic waste fragments in their seafood each year.

Thus, the question is how bad for us is the plastic we are eating in our seafood. Although we don't absorb very much plastic, it will accumulate in the body over time. And remember the food chain cycle: toxins are magnified at the top of the food chain. That means us.

GESAMP, a joint group of experts on the scientific aspects of marine environmental protection, confirmed that microplastic contamination has been recorded in thousands of organisms and more than 100 marine species. However, Richard Thompson, a leading international expert on microplastics and marine debris, claims that there is no evidence of harm to humans. But, he adds, "It's only going to increase. If we carry on with business as usual, it will be a different story down the line, in 10, 20 years. We're on the edge of a major ecological disaster."

From Sea to Soil

There's nothing worse smelling and looking than a pile of decomposing fish. It was, and sometimes still is, common practice to dump fish waste back into a lake or ocean. The trouble is that dumping it all in one place can overload the ecosystem, so dumping is banned to protect our bodies—marine, human, and water. Disposing of this waste can be a problem for anyone from big commercial food processors to small sport-fishing operations.

Some have found markets by grinding up the waste to make cat food or converting it to liquid fertilizer, but much of it is still going into landfills. In Alaska, where fish-related enterprises are the third-largest employer, three billion pounds of waste are generated annually. Several fish processors have taken advantage of the large amount of woodchips available from the local timber industry to operate large-scale fish and shellfish composting operations.

The FAO concluded that the loss of potential catch resulting from discarding fish waste around the world amounts to billions of dollars, and the economic losses due to bycatch (catch of nontargeted fish and ocean wildlife injured or killed in the process and considered waste) is equal or exceeds the value of targeted catch in some fisheries. As much as 2 to 10 billion pounds of fish are discarded by fisheries in the United States each year, hindering the recovery of some marine species.

There has also been an alarming decline in shark populations because of soaring demand for their fins, at $300 a pound, for a pricey soup that is considered a status symbol in some Chinese restaurants and Asian countries. Several states, among them California, have banned the sale or possession of shark fins, and shark sanctuaries are springing up around the world. Sharks are apex predators that play an important role in the ecosystem by maintaining the species below them in the food chain and serving as an indicator for ocean health. They help remove the weak, sick, and dead. And shark finning is one of the most egregious ways of squandering marine life on the planet, as the body of the shark is left for waste.

The World Bank estimated that poor management and overexploitation of fishery resources costs the global economy $50 billion every year, signaling that this is not just a problem in the United States. Despite notable improvements in a few specific regions and fisheries, industry practices have not significantly changed to ameliorate the monumental waste.

Catch These with Ease

American consumers now have the answer to sustainable fish shopping at their fingertips by using informative websites or reading *Bottomfeeder: How to Eat Ethically in a World of Vanishing Seafood*, by Taras Gresco. The Blue Ocean Institute monitors 90 seafood species for up-to-the-minute sustainability ratings and will text back either a safe code "green" or a danger code "red," for farmed and wild salmon, along with health advisory warnings about the possible presence of polychlorinated biphenyls (PCBs), dioxins, and pesticides. The EPA also offers warnings, especially to pregnant women, about various fish that contain dangerous levels of toxins, and other potentially dangerous waste. Greenpeace puts out an annual report called *CATO—Carting Away the Oceans*, and evaluates supermarkets on their buying and purchasing policies.

The same organization that provides ratings for Whole Foods is the FishPhone app, which includes a ranking system and suggestions on alternatives to overfished species. Safe Seafood provides rankings—"enjoy!" (green), "eat in moderation" (orange), and "avoid" (red)—for over 100 different types of seafood. Each species includes a photo, a description, and even a link to Wikipedia with more information about that type of seafood.

Precious Perishables

Of all the food that Americans waste, it's the seafood that never gets eaten that should trouble us most. Fish are high in protein, and low in fat. Eating them is associated with all sorts of health benefits. Unfortunately, because of their perishable nature, few foods are discarded so frequently. Between 2009 and 2013, as much as 47 percent of all edible seafood in the United States went to waste, according to a new study from the Johns Hopkins Center for a Livable Future. The majority of that is thanks to consumers who buy fresh and frozen fish—with good intentions— but never eat it. To put the scale of seafood loss in the United States in perspective, estimates suggest that the 2.3 billion pounds of seafood squandered each year would provide enough protein for more than 11 million people—for an entire year.

It seems the fate of fish even makes "Sorry Charlie the Tuna" seem appetizing.

SOURCES

"14 Food Waste Facts That Will Make You Want to Change the World." *Huffington Post*, February 7, 2017, https://www.huffingtonpost.com/foodbeast/14-food-waste-facts-that_b_4746413.html.

"21 Shocking Food Facts and Statistics and Ways to Reduce Waste." Visually, https://visual.ly/community/infographic/environment/21-shocking-us-food-waste-facts-statistics.

"10 Facts You Might Not Know about Food Waste." Foodtank, https://foodtank.com/news/2015/06/world-environment-day-10-facts-about-food-waste-from-bcfn/.

"All About: Food Waste." *CNN News*, January 22, 2008, http://www.cnn.com/2007/WORLD/asiapcf/09/24/food.leftovers/index.html.

"America's $165 Billion Food-Waste Problem." *CNBC*, July 17, 2015, https://www.cnbc.com/2015/04/22/americas-165-billion-food-waste-problem.html.

Arthur, Chris. "Food: Where Does It Really Go?" *Prezi*, December 12, 2014, https://prezi.com/kehfiilc6ykm/food-where-does-it-really-go/.

Best, James. "Hey, Farmers Market Snobs: Ugly Produce Needs Love Too." *Take Part*, May 27, 2014, http://www.takepart.com/article/2014/05/27/ugly-produce-needs-love-too.

Cambro. "Special Report: Food Storage and the Bottom Line." https://www.cambro.com/uploadedFiles/Content/General_Content/Custom_Pages/Resource_Center/ENG%20-%20Special%20Report%20AD.pdf.

"Carting Away the Oceans 2014." GreenPeace, May 12, 2014, https://www.greenpeace.org/usa/research/carting-away-the-oceans-2014/.

Curry, Maya. "The War on College Cafeteria Trays." *Time*, August 25, 2008, http://content.time.com/time/nation/article/0,8599,1834403,00.html.

Dewey, Caitlin, "You're about to See a Big Change to the Sell-By Dates on Food." *Washington Post*, February 16, 2017, https://www.washingtonpost.com/?utm_term=.51f0b1c9a822.

Dillinger, Jessica. "Largest Food Exports by Country." *World Atlas*, April 25, 2017, https://www.worldatlas.com/articles/the-american-food-giant-the-largest-exporter-of-food-in-the-world.html.

Eberlein, Sven. "Where No City Has Gone Before: San Francisco Will Be World's First Zero-Waste Town by 2020." *Alternet*, April 18, 2012, https://www.alternet.org/story/155039/where_no_city_has_gone_before%3A_san_francisco_will_be_world%27s_first_zero-waste_town_by_2020.

"Farm to Family: For Donors." California Association of Food Banks, http://www.cafoodbanks.org/farm-to-family-donor.

"Feeding America: Top 10 Facts about Hunger in U.S." Food Manufacturing, December 22, 2011, https://www.foodmanufacturing.com/news/2011/12/feeding-america-top-10-facts-about-hunger-us.

Ferdman, Roberto. "US Seafood Waste Enough to Feed up to 12 Million People for a Year." *Anchorage Daily News*, June 26, 2016, https://www.adn.com/nation-world/article/us-seafood-waste-enough-feed-12-million-people-year/2015/09/29/.

"Fighting Food Waste with Food Rescue." Feeding America, http://www.feedingamerica.org/our-work/our-approach/reduce-food-waste.html.

Firger, Jessica. "When It Comes to Food, Americans Are Shamefully Wasteful." *Newsweek*, July 21, 2016, http://www.newsweek.com/food-waste-80-billion-pounds-foodborne-illness-482849.

"Fish and Shellfish Advisories and Safe Eating Guidelines." U.S. Environmental Protection Agency, https://www.epa.gov/choose-fish-and-shellfish-wisely/fish-and-shellfish-advisories-and-safe-eating-guidelines.

"Food Waste." Food Left Over. https://foodleftover.weebly.com/food-waste1.html.

"Food Waste." Grace Communications, http://www.sustainabletable.org/5664/food-waste.

"Food Waste from Field to Table." Hearing before the Committee on Agriculture, House of Representatives, 114th Congress, Second Session, https://www.scribd.com/document/345738688/.

Fosdick, Dean. "Reducing Food Waste Is Good for the Earth and Your Wallet." *The Daily Courier*, February 10, 2015, https://www.dcourier.com/news/2017/feb/10/reducing-food-waste-good-earth-and-your-wallet/.

Gerlat, Allan. "Study Claims Most Americans Would Compost—If It's Easy and Cheap." *Waste 360*, January 8, 2014, http://www.waste360.com/composting-and-organic-waste/study-claims-most-americans-would-compost-if-it-s-easy-and-cheap.

Gruber, Karl. "Plastic in the Food Chain: Artificial Debris Found in Fish." *New Scientist*, September 25, 2015, https://www.newscientist.com/article/dn28242-plastic-in-the-food-chain-artificial-debris-found-in-fish/.

Gunders, Dana. "Wasted: How America Is Losing Up to 40 Percent of Its Food from Farm to Fork to Landfill." NRDC Issue Paper, August 2012), https://www.nrdc.org/sites/default/files/wasted-food-IP.pdf

Gunders, Dana. "Food Industry Moves to Remedy Expiration Dating Games." *Green Biz*, February 21, 2017, https://www.greenbiz.com/article/food-industry-moves-remedy-expiration-dating-games.

Gunders, Dana. "Super Size, Super Waste: What Whopping Portions Do to the Planet." *Grist*, October 15, 2012, https://grist.org/food/super-size-super-waste/.

Hausheer, Justine. "When Good Turkey Goes Bad: 10 Tips for Carving Out Food Waste." *Huffington Post*, January 20, 2013, https://www.huffingtonpost.com/onearth/when-good-turkey-goes-bad_b_2167875.html.

Hodges, Lauren. "A Brief History of Composting." Compost Pedallers, October 6, 2015, https://compostpedallers.com/compost/brief-history-composting.

Hradek, Christine. "The Rhoads' SNAP Challenge." Iowa State University, March 24, 2014, https://blogs.extension.iastate.edu/spendsmart/category/plan/page/6/.

Hsu, Tiffany. "A Powerful Use for Spoiled Food." *Los Angeles Times*, May 15, 2013, http://articles.latimes.com/2013/may/15/business/la-fi-ralphs-energy-20130516.

"Hunger in America." The Society of St. Andrew, http://endhunger.org/hunger-in-america/.

Hirsch, Jesse, and Reyhan Harmanci. "Food Waste the Next Food Revolution." *Modern Farmer*, September 30, 2013, https://modernfarmer.com/2013/09/next-food-revolution-youre-eating/.

Kelidjian, Amanda, Sara Young, Charlotte Grubb, and Dominique Cano-Stocco. *Wasted Cash: The Price of Waste in the U.S. Fishing Industry*. Oceana, 2014, http://oceana.org/sites/default/files/reports/wasted_cash_report_final.pdf.

Laskow, Sarah. "These Fish Are Eating the Plastic You Throw into the Ocean." *Grist*, November 4, 2013, https://grist.org/living/these-sea-creatures-are-eating-the-plastic-you-throw-into-the-ocean/.

Lieb, Emily B. et al. *Keeping Food Out of the Landfill*. Harvard Law and Policy Clinic, October 2016, http://www.endhunger.org/PDFs/2016/Harvard_FoodWaste_Toolkit_Oct2016.pdf.

Loki, Reynard, "America Produces a Shocking Amount of Garbage." *AlterNet*, July 14, 2016, https://www.alternet.org/environment/garbage-america-state-rankings-and-tips-reduce-waste.

Lott, Melissa C. "Why Eating Leftovers Is Good for Energy Efficiency." *Scientific American*, November 30, 2016, https://blogs.scientificamerican.com/plugged-in/why-eating-leftovers-is-good-for-energy-efficiency/.

Macdonald, Nancy. "We Throw Out, at Great Environmental Cost, a Horrific Amount of the Food We Grow." *McCleans*, November 9, 2009, http://www.macleans.ca/society/life/what-a-waste/.

Newcomer, Laura. "29 Smart and Easy Tips to Reduce Food Waste." *Greatest*, November 26, 2013, https://greatist.com/health/how-to-ways-reduce-food-waste.

Oliver, Rachael. "All about Food Waste." Food Waste in the World, https://foodwasteintheworld.weebly.com/index.html.

Peltz, Jennifer, and Bethan McKernan. "NYC Aims to Require Composting Food Scraps." *Global News*, June 18, 2013, https://globalnews.ca/news/649806/nyc-aims-to-require-composting-food-scraps/.

Plumer, Brad. "How the U.S. Manages to Waste $165 Billion in Food Each Year." *Washington Post*, August 22, 2012, https://www.washingtonpost.com/news/wonk/wp/2012/08/22/how-food-actually-gets-wasted-in-the-united-states/?noredirect=on&utm_term=.fb1267b0802c.

Reducing Wasted Food & Packaging: A Guide for Food Services and Restaurants. U.S. Environmental Protection Agency, https://www.epa.gov/sites/production/files/2015-08/documents/reducing_wasted_food_pkg_tool.pdf.

Salvador, Cesa Flavia, et al. "Synthetic Fibers as Microplastics in the Marine Environment." *Science of the Total Environment* 598 (November 15, 2017): 1116–29, Science Direct, https://doi.org/10.1016/j.scitotenv.2017.04.172.

"Senate Must Include 'America Gives More Act.'" in Tax Legislation, Food Bank of Contra Costa and Solano, & the *Vacaville Reporter*, October 14, 2014, https://www.foodbankccs.org/tag/charitable-giving.

Smillie, Susan. "From Sea to Plate: How Plastic Got into Our Fish." *Guardian*, February 14, 2017, https://www.theguardian.com/lifeandstyle/2017/feb/14/sea-to-plate-plastic-got-into-fish.

"So Much Waste." *River of Light*, 2014, http://www.riveroflightchicago.org/.

Steele, William. "Fish/Shellfish Waste Composting." Cornell Composting, http://compost.css.cornell.edu/fishwaste.html.

Szramiak-Arneberg, Camille. "Waste Not, Want Not: How Reducing Food Waste Can Help Address Climate Change." *Triple Pundit*, May 15, 2014, https://www.triplepundit.com/2014/05/waste-want-reducing-food-waste-can-help-address-climate-change/.

"The Number of U.S. Farmers Markets Has Nearly Doubled in the Last Five Years." *National Journal*, April 29, 2014, https://www.nationaljournal.com/s/58537/number-u-s-farmers-markets-has-nearly-doubled-last-five-years.

"U.S. Lets 141 Trillion Calories of Food Go to Waste Each Year." I Hate Working Retail, http://ihateworkinginretail.ooid.com/u-s-lets-141-trillion-calories-of-food-go-to-waste-each-year/.

Vince, Gaia. "How the World's Oceans Can Be Running Out of Fish." *Future*, September 21, 2012, http://www.bbc.com/future/story/20120920-are-we-running-out-of-fish.

"What Does McDonald's Do with Leftover Product at Closing Time?" *Quora*, May 20, 2015, https://www.quora.com/What-does-McDonalds-do-with-leftover-product-at-closing-time-.

Whitty, Julia. "The Fate of the Ocean." *Mother Jones*, March/April 2006, https://www.motherjones.com/politics/2006/03/fate-ocean/.

"Why They Matter." World Wild Life Fund, https://www.worldwildlife.org/species/shark#.

7

FASHION FROM TRENDY TO TRASH

Big-Time Blemishes of the Beauty Business

It's high irony that the industries that produce haute couture; flawlessly colored and perfectly cosmetized faces and hair; and produce impeccably lighted, posed, and glossy high-fashion photos, are also industries whose dirty secrets are hidden in clothes closets.

The fashion industry (including cosmetics) is the second-dirtiest business in the world, right after oil and petroleum products, with rampant production schedules, slick marketing, and unconscionable recycle rates. A lot of power is needed to produce 150 billion-plus articles of clothing each year, and most of the countries where those garments are produced use coal for their energy source. This helps to explain why the apparel industry is responsible for 10 percent of all carbon waste emissions globally. Clothing production is also an extremely land-and-water-intensive industry, and the apparel industry is one of the worst offenders when it comes to poor recycling, and introducing toxic chemicals into the worldwide water supply.

According to a 2013 report, cited by *Esquire* magazine, the global apparel industry produced enough garments in 2010 to provide 20 new articles of clothing for every person on the planet. By the way, in the 1960s, the American fashion industry produced 95 percent of its clothes in the United States, while by 2010 all but 3 percent were produced here. The incredible 500 percent increase in worldwide clothes production since the 1990s is due largely to the easy mark of teenagers and trendy "fast fashion."

We recklessly toss away more clothes than ever, largely because new clothes are getting cheaper, both to buy and to produce and sell. The United States generates 30.6 billion pounds of textile waste per year—enough to fill up two million Olympic-sized swimming pools. (Textile waste is material that is deemed unusable for its original purpose and created during fiber, textile, and clothing production, as well as consumer waste created during use and disposal.)

Today, the fashion industry and the culture of throwaway clothing that it has inspired have produced some startling statistics. The average American throws away more than 82 pounds of textiles per year—going directly into landfills. Across the globe, six million municipalities' ship textiles to landfills and incinerators. Because most of our clothing today is made with synthetic, petroleum-based fibers and aren't biodegradable, it will take decades, maybe centuries, for these garments to decompose in landfills, slowly releasing toxic vapors into the atmosphere. And the incineration of clothes made from polymers (plastics) foul the atmosphere with toxic chemicals.

We're buying more than 80 billion new items of clothing each year in this country, much of which is not being reused, recycled, or repurposed. "The amount of clothes and textiles being chucked away has been increasing steadily over the last 10 years and is sort of the dirty shadow of the fast fashion industry," states Christina Dean, founder and CEO of Redress, an organization focused on promoting environmental sustainability in fashion.

According to Secondary Materials and Recycled Textiles (SMART), a textile-recycling organization that partners with Goodwill and The Salvation Army, roughly 80 to 85 percent of our discarded clothes gets put in the trash and ends up in landfills, and 15 to 20 percent is donated, sold, recycled, given away, or are "taken back" by retailers.

Government data show that somehow Americans are throwing away even more of our unwanted duds than we are donating or recycling. One of the problems is that although there are people in need of clothing, there are far more clothing items than people who need them. And there are fewer items that are valuable enough to try to sell.

Charity is a nice sentiment, but it's grossly underrated as a way to get rid of clothes while helping someone else. Very little of all the clothing collected by charities and take-back programs are recycled into new textile fiber, according to H&M, a huge multinational clothing company.

One of the setbacks for donations is that different centers have diverse guidelines and standards. Some organizations accept used goods that are not in perfect condition, while others do not. As a result, 54 percent of respondents to a survey by the State of Reuse Report said they've thrown away an item because they didn't think it would be accepted by a donation center. The survey also found that 74 percent of people said they'd be more likely to donate old clothing if they knew that a center would take items regardless of condition.

According to the Council for Textile Recycling large organizations have well-developed systems for processing clothing. If items don't sell in the main retail store, they are sent to inexpensive outlets, where customers can walk out with a bag full of clothing for just a few bucks.

Vintage clothing stores around the world have been mining American thrift stores for cheap "era" fashion, which they then resell back to us at premium prices. Sharp shoppers snatch up the premium secondhand stuff at thrift store prices before they reach the vintage dealers. If not sold there, the clothing will be passed on to Goodwill's partners, who help divert two billion pounds of clothes from landfills annually through worldwide distribution and textile recycling.

The next stop for fashion castoffs is textile recyclers, just 1 percent of all clothing collected in such programs is recycled into textile fiber. Another 20 percent of the clothing—the ripped and stained items—is shipped out to processors that will chop it up into "shoddy," used in building insulation, carpet padding, or floor mats for the auto industry, the last gasp and cheapest type of clothing recycling.

America leads in the massive international export of secondhand clothing exchange, shipping $700 million worth of our unwanted apparel annually to other parts of the world—a fourfold increase in the last 15 years. Canada generally gets the best (especially the best winter stuff) and Japan gets the second-nicest vintage items. South American countries get the mid-grade stuff, Eastern European countries get the cold-weather clothes, and African countries get the stuff no one else wants. Now countries like China are refusing our hand-me-downs, and the low-grade clothing imports to sub-Saharan Africa are trashing their clothing industry.

Facts and figures:

- Estimates vary, but the global textile and garment industry (including textile, clothing, footwear, and luxury fashion) is currently worth $1.2 to $3 trillion worldwide. More than $250 billion is spent in America alone. In 2014, the average U.S. household spent an average $1,786 on apparel and related goods.
- Market value of accessories and footwear in the United States from 2011 to 2016 totaled almost $600 billion.
- The top fast-fashion retailers grew 9.7 percent per year over the last five years, topping the 6.8 percent of growth of traditional apparel companies, according to financial holding company CIT.
- Children's wear is expected to hit $173.6 billion by 2017.
- Each American throws away the equivalent of 200 T-shirts a year. Less than 15 percent of those are recycled.
- The Environmental Protection Agency (EPA) estimates that removing toxic textiles would be the equivalent of taking almost 7.5 million cars off the road.
- In order to manufacture nylon, nitrous oxide is released as part of the process—a greenhouse gas that is 32 times more potent than carbon dioxide.
- The manufacture of synthetic fabrics requires large amounts of crude oil, releasing toxic compounds when used, aggravating respiratory diseases, and contributing to climate change.

Donations or the Dump

When keeping up with the trends of fashionistas, the EPA estimates that the textile industry recycles approximately 3.8 billion pounds of postconsumer textile waste (PCTW) each year. But it only accounts for approximately 15 percent of all PCTW, leaving the bulk, dumped in our landfills.

Domestic resale has boomed in the era of the internet as consignment and thrift shop sales are growing at 5 percent per year, according to the National Association of Resale and Thrift Shops, which put Goodwill's sales of donated goods at thrift shops at around $4.6 billion annually.

The Association of Resale Professionals found that about 12 to 15 percent of Americans shop at consignment or resale stores. The Council for Textile Recycling estimates that 2.5 billion pounds of post consumer textile waste is collected and prevented from entering directly into the waste stream. This represents 10 pounds for every person in the United States, but it is still only about 15 percent of the clothing that is discarded.

Charities sell it to approximately 3,000 textile recyclers at five to seven cents per pound. One company processes more than 12 million pounds of postconsumer textiles per year. At these facilities, workers separate used clothing into 300 different categories by type, size, and fiber content. About 45 percent of these textiles continue their life as clothing sold in more than 100 countries.

Fast Fashion

Enter the disrupters—where dresses can be bought for under $10, clothes are manufactured quickly and inexpensively, and fashion choices are offered not in seasons but in weeks. It's called "fast fashion," a term used to describe the quick process in which designs move rapidly from catwalk to retailers so consumers can get what they want "right now."

It is described as "low-cost clothing collections that mimic current fashion trends." The garments are usually made out of lower-quality materials that are bought mostly by young consumers who crave to look in style, but on a budget. Because of this trend, the United States has produced the most gluttonish consumption and waste clothing problem in the world.

Fashion fiends love the idea of going into a clothes outlet and coming out with designer labels at a fraction of the price. Unfortunately, they don't really get clothes with "real" designer labels. Clothes outlet brokers deal with designers so they can put designer labels on the cheaply made clothing manufactured in their own low-quality factories and sell it for cheap.

Check out your "waste" size. "I think after any big change in any industry it takes a while to feel and smell the dirt that comes out of something that is polluting," says Christina Dean, founder and CEO of

Redress, an organization focused on promoting environmental sustainability in fashion.

Fast-Fashion Followers

One of the most coveted creatures in the trade is the teenage wannabe fashion maven leafing through sleek magazines, deciding what she wants to buy and what her friends think is trending. Master marketers of fast fashion are selling the sizzle, not the steak. In other words, perception is as powerful as reality, especially in selling clothes. It's the image, and how we look and feel about ourselves, and, of course, snake oil marketing.

Fast fashion has also been criticized for encouraging a "throw-away" attitude among consumers. Retailers like H&M, a billion-dollar Swedish company; Forever 21, a multibillion-dollar U.S. company; Zara, a Spanish company and currently the world's largest apparel retailer, which made its founder one of the wealthiest men in the world; Target, the second-largest discount store retailer in the United States; Walmart, the largest retailer in the world; and Amazon, the fastest-growing retailer in the world keep us buying a continuous supply pile of cheap, speedy, available, and trendy styles.

Some big outlets are restocking shelves with new merchandise on a weekly basis, moving about 80,000 pounds of clothing a day. The top fast-fashion retailers grew 9.7 percent per year over the last five years, topping the 6.8 percent growth of traditional apparel companies.

Clothing companies are wielding a double-edged razor. They know that fashion is getting out of hand, yet they don't want to stop selling their products or give up on their fast-fashion business models—it's been extraordinarily lucrative.

Fashionably Polluting the Planet

Chemicals, such as formaldehyde, perfluorinated chemicals (PFC), nonylphenol ethoxylate (NPE), p-Phenylenediamine (PPD), and dioxin-producing bleach are volatile organic compounds (VOCs), and all are commonly found and linger in our clothing. These chemicals have been shown to produce seriously adverse health effects. Some popular fast-fashion outlets are still selling lead-contaminated purses, belts, and shoes, years after signing a settlement agreeing to limit the use of heavy metals in their products.

Worldwide Patterns of Waste

That dress you just bought might be better traveled than you are. Raw materials can be shipped from the United States to places like Bangladesh, Vietnam, Pakistan, and the Philippines where they are used to manufacture clothing. The garments are then sent by rail, container ship, and truck to eventually end up at the retailer. There's no way to gauge how much fuel is used in total, but the fast-fashion industry's contribution to petrochemical emissions into the biosphere are big-time.

Americans purchase approximately a billion garments made in China each year, the equivalent of four pieces of clothing for every U.S. citizen, according to Pietra Rivoli, author of *The Travels of a T-Shirt in the Global Economy*. Once bought, an estimated 21 percent of annual clothing purchases stay in the closet, representing a potentially large quantity of latent leftovers that will eventually enter the solid waste stream, according to the EPA.

Cotton

Cotton is one of the thirstiest crops in existence, and yet the federal government subsidizes growing cotton in the desiccated land of Arizona, made possible by importing billions of gallons of water each year. Yet natural fibers have largely been substituted for synthetic ones—fabrics made from oil. The manufacture of polyester textiles soared from 5.8 million tons in 1980 to 34 million in 1997 to an estimated 100 million in 2015.

In theory, cotton is biodegradable and polyester is not. But the way we dispose of clothing makes that irrelevant. For cotton clothes to break down, they have to be composted, which doesn't happen in a landfill.

A comprehensive study of various textiles found that environmental impacts per kilogram of fiber are higher for cotton than other materials, primarily because of the large quantities of toxic pesticides and fertilizers required to grow cotton.

Polyester

Since 1951, with the first polyester suits made from fabric created by DuPont, it has been a fixture in our closets. It's already ubiquitous in our most basic garments. It's inexpensive, easy to blend with other materials, remarkably improved in its look and feel, and maybe no worse for the environment than conventionally grown cotton.

There are a number of synthetic options, such as rayon and nylon, but the preferred alternative—because it's cheap—is polyester. Tecnon Orbichem, a company providing data and analysis to the petrochemical industry, estimates that more than 98 percent of future fiber production will be synthetics, and 95 percent of that synthetic fiber will be polyester.

Even though it can be made from old plastic bottles, allowing companies to turn garbage into clothing, its biggest drawback is that it requires a lot of energy, which means burning fuel for power (again, mostly fossil fuels), contributes to climate change.

Zero Fashion

Sustainable fashion refers to items of clothing that generate little or no textile waste in their production. They can be divided into two general approaches: preconsumer zero-waste, which eliminates waste during manufacture, and postconsumer zero-waste, which generates material from secondhand clothing, eliminating waste at the end of product use.

Recycling is a solution to intercepting waste before it becomes an environmental hazard. What is needed are innovative processes in which solvents are used to selectively dissolve different types of textiles, recapturing them as a raw material, which can be used to make new clothes.

Several clothing retailers have announced take-back programs that collect used garments from customers to be recycled, sold, or remade into clothing. This plays into the crucial concept of extended producer responsibility—considering the product's afterlife. Some fashion businesses use fabric waste in a process referred to as "upcycling" rather than recycling.

A 2015 study released by Nielsen says that 66 percent of respondents, aged 15 to 20, are willing to pay more for products and services from socially and environmentally committed companies, an increase up from 55 percent from 2013.

"Haulternatives"

A year or two ago, a few users began uploading YouTube videos of themselves exchanging clothes with friends and showcasing how they made new styles out of old instead of tossing them out, or constantly buying new clothes. One YouTuber from England made a "Haulternative" video in which she was exchanging clothes with another YouTuber from Texas, looking at how people can refresh their wardrobe without having to buy new. Some companies are experimenting with new ideas like Rent The Runway, renting clothes to customers who pay a monthly fee.

Fashion by "Trashion"

It's a cute mating of "trash" with "fashion" and a term for art, jewelry, fashion, and object d'art created from used, thrown-out, found, and repurposed materials. It's perhaps the trendiest use of throwaways. Refashioning finds and generates value in cast-offs, making something from nothing for aesthetic purposes, and calls attention to waste and how it can be repurposed.

MIT sponsors a Trashion fashion show celebrating creative fashion design while promoting waste reduction and sustainability. Sonoma County, in Northern California, hosts a yearly trashion show. "So much of what gets thrown away on a daily basis can be reused or repurposed into something remarkable," said Toni Castrone, the community center's executive director. Trashion fashion even has a Facebook page (www.facebook.com/trashionfashionshow), and there are thousands of photos on Flickr websites, scads of designers, followers, and Trashion shows worldwide.

COSMETICS: THE UGLY SIDE OF PRETTY

Despite our best intentions, the dozens of bottles and tubes anointing our bathroom shelves are mostly destined for landfills. And some of these unguents and anointments to enhance beauty can be unhealthy. Probing into the world of lotions and potions that promise to de-wrinkle, soften, illuminate, and glossify, it becomes apparent that this market is creating a lot of ugly waste in the name of beauty.

Even the most ubiquitous personal care products, toothbrushes and razors, for example, pose an environmental headache. Around 23,000 tons of toothbrushes end up in U.S. landfills every year (a figure that would be higher if more people changed their toothbrushes every three months as recommended), and each year, two billion disposable razors are thrown away. The razor company BIC introduced the BIC ECOlutions Shaver, a supposedly "green" disposable razor that utilizes bioplastics as a hopeful solution. Unfortunately, bioplastics are a hassle to recycle, as are other plastic disposable razors, one problem is because of their size.

One toiletries company saves more than 450 tons of virgin plastic each year partially by recycling 37 million plastic caps. A natural skin-care specialist has a program inviting customers to return packaging in exchange for free product samples. So far, over seven tons of cosmetic packaging from that one company has been saved in North America alone.

Another company striving to eradicate the environmental damage that looking good can cause is trying to get away with minimal packaging, consumer-returns incentives, and highly recyclable packaging. It's able to manage the entire cradle-to-grave process because they use the cheapest, safest, lightest packaging they can find, or use no packaging at all.

Cool Cosmetology

A half billion "selfies" can't be wrong. It's been suggested that because young women have a mind-set of being "selfie-ready," lip and eye products have become two of the best-performing cosmetic categories for the last three years, tracking "almost exactly" with the rise of smartphones and mobile devices.

Cosmetics from foundation to lip gloss to eye shadow have increased in sales by something like 20 percent over the last few years, and by 2022, the American cosmetic industry will represent more than $58.8 billion in sales. Beauty has been the fastest-growing retail segment for at least the last three years, outpacing apparel, footwear, even food. Unlike other household items such as kitchenware or clothes, cosmetics can't be donated or recycled after they have been used—all women can do is throw them away—safely one hopes. If various cosmetic potions contain truly noxious or irritating ingredients (look to see if they're listed as toxic on the Environmental Working Group's Skin Deep database), do not dump them down the drain or toilet.

Dump out a woman's handbag and you might find a 10-year-old lipstick, a 5-year-old tube of mascara, and a jar of rancid moisturizer, all teeming with harmful bacteria. "It's ironic that women spend so much time and money on cosmetics to make themselves look beautiful, and yet by using it, they are making themselves vulnerable to infections that are anything but attractive," said Dr. Susan Blakeney, an English optometrist specializing in eye infections.

Use-By Dates

Unlike food, the Food and Drug Administration (FDA) doesn't really have tough rules and regs concerning most cosmetic products. Products are not required to have an expiration date on their packaging, and any guidelines are usually offered by the cosmetics industry. The FDA can issue advice and warnings about misleading claims and the use of color additives that aren't in accordance with the FDA's guidelines, but otherwise the consumer is on her own.

Important considerations when deciding when to toss and replace beauty products are the type of makeup, how it's been stored, and whether you've lately had an eye or skin infection. If a thin, watery layer appears on any makeup surface, they should be discarded immediately. It is best to be on the side of caution and save yourself a visit to the doctor.

When you open a cosmetic for the first time, write the date on the product. It will help you keep track of how long you've had the product, so you'll know when it's time to throw it out. To get maximum longevity from your makeup, store it in a cool, dry place, away from direct sunlight. Always close the lids tightly after use.

Mascara should be thrown away after three to six months—it has the shortest life span of all makeup because of the risk of transferring bacteria from your eye to the mascara tube, which can drive bacteria deep down into the product waiting to invade the eye. Over a third of women under 24 admit to sharing mascara with friends, accounting for a growing eye infection problem. Never share makeup, especially lipstick, mascara, eye shadow, or pencils. If your lipstick smells even the slightest bit rancid, pitch it.

Moisturizing foundations and stick concealers are good for around 12 to 18 months. If the product has separated, looks chalky, has an "off" smell, or has entirely changed in color, lose it right away. Use clean brushes or sponge applicators to apply the product to prevent bacterial contamination. Be sure to wash or replace your applicators frequently. Eye and lip pencils will keep for 18 months. Pressed, loose, or shimmer powder, eye shadow, blush, lip gloss, lipstick, and toner are good for about a year.

To clean eye shadow, powder blush, lip liners, eye pencils and lipsticks—if the container, lid, and packaging is airtight and clean—wipe the makeup with a cotton swab or cotton pad soaked in isopropyl alcohol. This will kill any bacteria without affecting the shade and texture of the product.

Empty First, Toss Later

As you clean out your makeup drawer, you're likely to wind up with plenty of containers with partial remnants. While your first instinct may to be toss them in the trash, or to rinse out and recycle these bottles, jars and tubes, doing so sends these toxic waste chemicals right into the landfill or water table and conventional waste water treatment plants break down only about 10 to 12 percent of personal and pharmaceutical care products (PPCPs).

Pour the contents from each rejected bottle into one container and then recycle the rest. Call or go online to find out if your local disposal center accepts cosmetics as household hazardous waste. Bringing them there ensures they'll be disposed of safely.

There are compelling reasons for tossing out all "non-green" products commonly found in cosmetics, such as parabens (preservatives sometimes affecting estrogen production or other aspects of the endocrine system), which are linked to a heightened risk of breast cancer; and synthetic musks, which have been detected in human breast milk. When the Women's Environmental Network (wen.org.uk) tested toiletries, it found estrogen-mimicking preservatives in nearly 60 percent of cosmetic products. And they're not good for aquatic life either, so keep them out of the waterways.

Contact with Contacts

Do not flush your contacts down the toilet or rinse them down the sink, as that is a one-way ticket to local rivers and streams. Most cannot be recycled and none of them belong cluttering up our waterways or in the intestines of marine animals.

Barefaced Can Be Beautiful

No health studies, premarket testing, or government regulations are required for cosmetic products, according to the Office of Cosmetics and Colors at the FDA. Many of these products can be counterfeit and come from countries that have absolutely no regulation or conscience, like China. Remember that beauty is more than skin deep—no matter what you use on your skin.

SOURCES

"16 Facts about Cotton That You Don't Know." Barnhardt Purified Cotton, April 26, 2016, https://www.barnhardtcotton.net/blog/16-facts-about-cotton-that-you-dont-know/.

Au-Yeung, Angel. "Forever 21 on Forbes Lists." *Forbes*, https://www.forbes.com/companies/forever-21/.

Bain, Marc. "Polyester: The New Fabric of Our Lives." International Cotton Advisory Committee, June 5, 2015, https://www.icac.org/getattachment/cmte-cotton-industry/Private-Sector-Advisory-Panel-PSAP/Documents/PSAP-34-Att-1-Polyester-the-new-fabric.pdf.

"Bic Disposable Razor." *Plastic News Europe*, November 19, 2009, http://www.plasticsnewseurope.com/article/20091119/PNE/311199986/bic-disposable-razor-1975.

Blandchard, Tamsin. "Meet the YouTubers Making #Haulternative Videos for Fashion Revolution Day." *Telegraph*, April 24, 2011, http://fashion.telegraph.co.uk/news-features/TMG11536302/Meet-the-YouTubers-making-haulternative-videos-for-Fashion-Revolution-Day.html.

Breyer, Melissa. "25 Shocking Fashion Industry Statistics." *TreeHugger*, September 11, 2012, https://www.treehugger.com/sustainable-fashion/25-shocking-fashion-industry-statistics.html.

Bryant, Kelly. "You Won't Believe How Much Clothing the U.S. Throws Away in a Year." *Take Part*, May 29, 2015, http://www.takepart.com/video/2015/05/29/clothes-trash-landfill.

Bryson, Douglas, and Glyn Atwal. "Responsible Luxury." In *Luxury Brands in China and India*. London: Palgrave Macmillan, 2017, https://link.springer.com/chapter/10.1057%2F978-1-137-54715-6_8.

Burgess, Lindsay. "6 Items You Should Never Share with Your Roommate (or Anyone)." *Brit+Co*, January 18, 2018, https://www.brit.co/health-risks-of-sharing/.

Christian, Scott. "Fast Fashion Is Absolutely Destroying the Planet, and Buying a Prius Won't Make Up for It." *Esquire*, November 14, 2016, https://www.esquire.com/style/news/a50655/fast-fashion-environment/.

Claudio, Luz. "Waste Couture: Environmental Impact of the Clothing Industry." *Environmental Health Perspectives* 115, no. 9, September 2007, https://www.ncbi.nlm.nih.gov/pmc/articles/PMC1964887/.

Cline, Elizabeth. "America Leads the World in Textile Waste and Unwanted Clothing, Here's Why." Pratt, July 26, 2016, https://bkaccelerator.com/america-leads-world-textile-waste-unwanted-clothing/.

Cline, Elizabeth. "The Afterlife of Cheap Clothes." *Slate*, June 18, 2012, http://www.slate.com/articles/life/fashion/2012/06/the_salvation_army_and_goodwill_inside_the_places_your_clothes_go_when_you_donate_them_.html.

D'Angelo, Stephen. "Team Develops Machine with Aim of Ending Textile Waste." *Cornell Chronicle*, April 19, 2017, http://news.cornell.edu/stories/2017/04/team-develops-machine-aim-ending-textile-waste.

Densmore, Sarah. "How Long Should You Keep Your Make Up?" Dummies, http://www.dummies.com/health/how-long-should-you-keep-your-make-up/.

Dwyer, Liz. "Here's Why We Trash 26 Billion Pounds of Clothing a Year." *Take Part*, June 10, 2016, http://www.takepart.com/article/2016/06/10/why-we-trash-26-billion-pounds-clothing-year.

"Ecochic Design Award: Sourcing Textile Waste." Redress, 2014, https://static1.squarespace.com/static/582d0d16440243165eb756db/t/585a15a9bebafba69927c172/1482298805626/LEARN2014_Sourcing_ENG-07.pdf.

England, Rachel. "Wasted Beauty: Packaging in the Cosmetics Industry." *Resource*, November 16, 2010, https://resource.co/article/Packaging/Wasted_beauty_Packaging_cosmetics_industry.

Farag, Nadine. "Know Your Materials: What Each One Means for Sustainable Fashion." *Man Repeller*, June 27, 2016, https://www.manrepeller.com/2016/06/sustainable-fashion-materials.html.

Fassa, Paul. "Chemical Clothing: Which Hidden Toxins Are You Wearing?" *Natural Society*, April 8, 2013, http://naturalsociety.com/chemical-clothing-toxic-chemicals-clothes-sick/.

"Fast Fashion." The Fashion Law, October 3, 2016, http://www.thefashionlaw.com/learn/fast-fashions-green-initiatives-dont-believe-the-hype.

"Fast Fashion." Wikipedia, https://en.wikipedia.org/wiki/Fast_fashion.

"Fast Fashion Is Creating an Environmental Crisis." *Green Stitched*, September 6, 2016, https://greenstitched.com/2016/09/06/1608/.

Gensle, Lauren. "The World's Largest Retailers 2016: Wal-Mart Dominates but Amazon Is Catching Up." *Forbes*, May 27, 2016, https://www.forbes.com/sites/laurengensler/2016/05/27/global-2000-worlds-largest-retailers/#220ed0c6bbb0.

"Global Fashion Industry Statistics—International Apparel." Fashion United, https://fashionunited.com/global-fashion-industry-statistics.

Hill, Amelia. "Beauty Experts Slap Health Warning on Lipsticks Past Their Sell-By Date." *Guardian*, June 3, 2007, https://www.theguardian.com/uk/2007/jun/03/fashion.lifeandhealth.

Hochwald, Lambeth. "When to Throw Away Your Makeup." *Mother Nature Network*, April 21, 2016, https://www.mnn.com/lifestyle/natural-beauty-fashion/stories/when-throw-away-your-makeup.

"How H&M Is Trying to Balance Fast Fashion with Revolutionary Recycling." *Green Stitched*, December 31, 2016, https://greenstitched.com/tag/closed-loop-textile-recycling/.

"If Your Clothes Aren't Already Made out of Plastic, They Will Be." Quartz, https://qz.com/414223/if-your-clothes-arent-already-made-out-of-plastic-they-will-be/.

Kronstad, Sylvia. "Thrift Shop: It's Not Just a Song—It's a SCAndal." *The Economic Populist*, July 4, 2010, http://www.economicpopulist.org/content/thrift-shop-its-not-just-song-its-scandal-5552.

Lagosi, Louise. "EcoSalon Investigates: What Happens to Our Cast Off Clothing?" *EcoSalon*, May 27, 2011, http://ecosalon.com/ecosalon-investigates-what-happens-to-our-cast-off-clothing/.

Lake, Rebecca. "23 Teenage Consumer Spending Statistics That Will Shock You." Credit Donkey, December 15, 2014, https://www.creditdonkey.com/teenage-consumer-spending-statistic.html.

Lane, Kara. "Why Smart People Don't Wear Fast Fashion or High-End Designer Clothes." https://karalane.com/smart-people-dont-wear-fast-fashion-high-end-designer-clothes.

LaRose, Danielle. "A Step Too Far: Fashion's Carbon Footprint." Carman Busquets, April 20, 2017, https://www.carmenbusquets.com/journal/post/fashions-carbon-footprint.

Lewis, Kevin. "The Fake and the Fatal: The Consequences of Counterfeits." *The Park Place Economist* 17, 2008, 47–58, https://www.iwu.edu/economics/PPE17/lewis.pdf.

Low, Elaine. "Beauty CEO Amin on Social Cosmetics and the Ageless Beauty Boom." *Investor's Business Daily*, December 29, 2016, https://www.investors.com/news/e-l-f-beauty-ceo-amin-on-social-cosmetics-and-the-ageless-beauty-boom/.

Lustgarten, Abrahm, and Naveena Sadasivam. "Why Are Federal Dollars Financing This Thirsty Crop in Arizona?" *Grist*, May 27, 2015, https://grist.org/climate-energy/why-are-federal-dollars-financing-this-thirsty-crop-in-arizona/.

Mikashen1007. "Research Zero Waste Design," *Textile Innovation* (blog), November 13, 2014, https://liangyinshen07.wordpress.com/2014/11/13/research-zero-waste-design/.

Mitchell, Katalina. "Makeup 101: How Long Your Fave Products Really Last! Find Out When It's Time to Toss Your Makeup, Perfume, Nail Polish and More." *Seventeen*, December 12, 2012, https://www.seventeen.com/beauty/makeup-skincare/advice/g370/makeup-expiration-date-guidelines/.

Rabbitt, Emily. "2 Billion Tossed Per Year: What's the Most Wasteful Bathroom Product?" Groundswell Stories, October 29, 2014, https://groundswell.org/2-billion-tossed-per-year-whats-the-most-wasteful-bathroom-product/.

"Resale Thrives in Slow Economy." Narts, July 8, 2015, https://www.narts.org/i4a/pages/index.cfm?pageid=3290.

Ruddick, Graham. "How Zara Became the World's Biggest Fashion Retailer." *Telegraph*, October 20, 2014, https://www.telegraph.co.uk/finance/newsbysector/retailandconsumer/11172562/How-Inditex-became-the-worlds-biggest-fashion-retailer.html.

Salkin, Allen. "Who Killed Paper Clothing? A Modest Proposal on How Clothing Might Be Reborn." *Good*, no. 35, December 28, 2016, https://www.good.is/features/issue-35-who-killed-paper-clothing.

"Shelf Life and Expiration Dates: The Science & Safety behind Your Favorite Product." Cosmetics Info, http://www.cosmeticsinfo.org/shelf-life-and-expiration-dates.

Shieve, Tom. "Are Parabens Dangerous." How Stuff Works, https://health.howstuffworks.com/skin-care/beauty/skin-and-lifestyle/parabens1.htm.

Siegle, Lucy. "How Should I Dispose of Old Cosmetics?" *Guardian*, August 30, 2008, https://www.theguardian.com/environment/2008/aug/31/ethicalliving.beauty.

Soo Hoo, Nia. "The Comprehensive List of Where You Can Return Opened Beauty Products." Fashionista, August 20, 2015, https://fashionista.com/2015/08/where-to-return-opened-beauty-products.

Sweeny, Glynis. "Fast Fashion Is the Second Dirtiest Industry in the World, Next to Big Oil." EcoWatch, August 17, 2015, https://www.ecowatch.com/fast-fashion-is-the-second-dirtiest-industry-in-the-world-next-to-big--1882083445.html.

Sweeny, Glynis. "It's the Second Dirtiest Thing in the World—and You're Wearing It." *AlterNet*, August 13, 2015, https://www.alternet.org/environment/its-second-dirtiest-thing-world-and-youre-wearing-it.

Tan, Judith. "Singapore: Throw-Away Nation." *Business Times*, January 15, 2017, http://www.straitstimes.com/singapore/singapore-throw-away-nation.

Telfer, Tori. "When to Throw Away Old Makeup and Why You Should." *Bustle*, November 19, 2013, https://www.bustle.com/articles/9165-when-to-throw-away-old-makeup-and-why-you-should

"Textile Recycling Facts and Figures." *The Balance Small Business*, December 29, 2017, https://www.thebalancesmb.com/textile-recycling-facts-and-figures-2878122.

"This Is Exactly What Happens When You Use Old Makeup on Your Face." *Best Health*, http://www.besthealthmag.ca/best-looks/throw-out-old-makeup/.

"Trashion Fashion." *Sonoma Index Tribune*, March 27, 2015, http://www.sonomanews.com/entertainment/3714274-181/trashion-fashion?sba=AAS.

Umbra. "What's the Best Way to Dump My Old Cosmetics and Contact Lenses?" *Grist*, July 9 2015, https://grist.org/living/whats-the-best-way-to-dump-my-old-cosmetics-and-contact-lenses/.

"Waste & Recycling." The Source, January 2010, http://source.ethicalfashionforum.com/article/waste-recycling.

"Waste: When the Stuff You Throw Away Doesn't Really Go Away." Canopy, https://weare.tearfund.org/wp-content/uploads/2017/11/Cause-Waste.pdf.

"Why This Matters—Cosmetics and Your Health." EWG Cosmetic Database, https://www.ewg.org/skindeep/2011/04/12/why-this-matters/#.WzP8N9VKiUk.

Wicker, Alden. "Fast Fashion Is Creating an Environmental Crisis." *Newsweek*, September 9, 2016, http://www.newsweek.com/2016/09/09/old-clothes-fashion-waste-crisis-494824.html.

Wiranda, Timothy. "Tag: Waste Reduction." makeasmartcity, October 26, 2017, https://makeasmartcity.com/tag/waste-reduction/.

Yuntan, Zhia. "What Happens When Fashion Becomes Fast, Disposable and Cheap?" NPR, April 10, 2016, https://www.npr.org/2016/04/08/473513620/what-happens-when-fashion-becomes-fast-disposable-and-cheap.

Zhu, Yehong. "Here's Why You Shouldn't Freak Out over H&M's Sales Growth Slowdown." *Forbes*, June 23, 2016, https://www.forbes.com/sites/yehongzhu/2016/06/23/heres-why-you-shouldnt-freak-out-over-hms-sales-growth-slowdown/#36c77a64206b.

8

SWEET AND SOUR CHARITY

Charity is a battle of the head and the heart, generosity and greed.
—*Anonymous*

"Are there no prisons?" asked Scrooge. And the union workhouses?" he demanded. "I help to support the establishments I have mentioned. They cost enough and those who are badly off must go there."

"Many can't go there; and many would rather die," said one gentleman.

"If they would rather die," said Scrooge, "they had better do it, and thus decrease the surplus population."
—*Charles Dickens, "A Christmas Carol"*

Depending on the time of the year, the place, the people, and frame of mind, one might want to give generously or not at all. It's a matter of conscience. If you feel benevolence seeping into your soul, this chapter will help you to be careful about how, what, and to whom you choose to contribute. There is much need, but there is also a lot of waste and cunning conning cloaked in an aura of goodwill. Therefore, be aware that there are those who prey upon the good works of others, and be watchful lest good intentions go awry.

SCROOGE OR STOOGE?

There are those who claim that charity is no solution to poverty, that it only perpetuates it by stripping the ill fated of initiative, and only serves to appease and ease our own conscience. Others claim that it is the most selfless of actions—if done unselfishly. Most of us land somewhere in the middle, between the miserly misanthrope and Christianity's most compassionate prophet. When it comes to discarding things we no longer want or need, or to part with some capital, charity can be a good option, but the impulse to give can also have deleterious effects as well as leave us vulnerable to con artists. This chapter is not a shot aimed at charities themselves or those who wish to give to others, but a cautionary tale about what altruism is about psychologically and physiologically, what makes us want to give, and what to look for to avoid being taken for a "mark."

WHAT YOU DON'T KNOW ABOUT WHY YOU HELP OTHERS

There is no doubt that philanthropy has helped a lot of people and animals through some very tough times. People say that charity is its own reward—but there are other bonuses of which you should be aware: social status, feeling a warm glow, rationalization, and getting high. Yes, charity can also get you high. Literally. It affects the area of the brain that secretes powerful chemicals—the "feel good stuff"—like dopamine, serotonin, oxytocin, and endorphins. These are the chemicals that control the brain's reward, pleasure, and pain-killing centers. Our brain enables us not only to feel rewards, but also to take action to get them.

People also take pleasure in charitable giving because it enhances their social status. They enjoy being regarded as generous and as possessors of disposable wealth, especially when publicly distinguished with titles like "sponsor" or "patron." The Irish author George Bernard Shaw quipped that a rich man, "does not really care whether his money does good or not, provided he finds his conscience eased and his social status improved by giving it away."

In giving, we might even be only thinking about our own future. For example, when people donate to a university, perhaps they expect their child to study there. When they donate to a hospital, they may be thinking about the day they need its services.

Finally, Ernest Becker, in his Pulitzer Prize–winning book, *The Denial of Death*, posits that humans try to give their life a meaning that will carry on after death. The "tithes" we offer can go on doing good in our name.

Tricks for Treats

The trick in seeing a hungry child or a scared animal is a phenomenon well known to fund-raisers called the "identifiable victim effect." This is the human inclination to be more helpful to actual individuals—with a name or a face—than to anonymous victims. This may be a tendency to pitch in to ensure the survival of those with whom we identify closely. This rubs up against what is called "futility thinking" or the idea that giving money to help the poor is a waste, like focusing on those we cannot save rather than on those we can.

Hey, it's OK to catch a pleasant buzz from a righteous and benevolent act. And it's worth observing that reasons for virtuous behavior have been a topic of interest in many philosophical and religious traditions. Mahatma Gandhi once said, "The best way to find yourself is to lose yourself in the service of others." The Buddhist concept of dāna, or pure altruism, is when giving is divorced from any internal reward and is a key attribute to enlightenment. Generosity is one of the 10 pārāmitas in Hindu religious texts, the enlightened quality of giving.

Healthy and Happy

Still, some ask, why we should give at all, isn't it a waste of my time and money? Don't we have a right to keep our hard-earned money for ourselves? Perhaps, but a survey of 30,000 American households, reported in *Newsweek,* found that those who gave to charity were 43 percent more likely to say they were "very happy" about their lives than those who did not. Of those who volunteer time, 68 percent reported that it made them feel physically healthier, 89 percent said that it "has improved my sense of well-bring, and fulfillment" and 73 percent claimed that it "lowered stress levels."

In 2008, researchers from Harvard gave participants money in a study. They asked half to spend it on themselves, and the other half to give it to some person or charity. Those who donated the money showed a significant uptick in self-esteem and happiness; those who spent it on themselves did not.

Philanthropy Is Undemocratic

Whether on the right, the left, or in the middle, a shared point of view seems to be that "fat cat" donors accumulate too much power about how much should be allocated to whom, and for what, imposing their will on society. Only 14 percent of total giving today comes from foundations started by the wealthy. The vast majority of annual donations are by everyday donors who give at a rate of about $2,500 per household, according to the Stanford Social Innovation Review.

Alexis de Tocqueville wrote in *Democracy in America* that a healthy democracy should encourage, not discourage, people from acting on their own to improve the common good. The millions of givers compose a model that sets goals and priorities, and collects resources to take on problems without getting involved with the government. This choice and decision-making would be impossible if we believed that only the state should make these decisions.

Then there are those who point out that a good portion of our budget through taxes goes toward funding philanthropic causes. We don't have control over that money, which, some feel can be wrongly allocated. Plus, it's difficult to impossible at times to know if a charity is corrupt. People waste billions of dollars on inefficient, poorly run, or downright fraudulent charities because they can't or do not bother to research where their money is going.

In a nationally representative survey in 2015, respondents chose private philanthropy over government as their "first choice for solving a social problem in America," by 47 percent to 32 percent. Asked whether they most trusted entrepreneurial companies, nonprofit charities, or government agencies, 43 percent of respondents chose charities, 28 percent selected entrepreneurial companies, and just 14 percent chose government agencies. Personal philanthropy tends to be more inventive and experimental, quicker, more efficient, and varied.

THE BOTTOM OF THE BENEVOLENCE BARREL

Charity is sometimes like a magic act. What you think you see is largely sleight of hand. Many charities use popular causes such as armed service veterans, firefighters, children's causes, and animals with beautiful fur and large eyes as shills. They work on people to get an emotional response and then a donation.

"Causes" such as the Cancer Fund of America bilked millions from donors and spent less than 3 percent of it on cancer patients. The rest went to themselves, family, and friends. Donald Trump and his son Eric have also been accused of slippery dealings with their "charitable" foundation that funds cancer research for kids, according to *Forbes* magazine.

Some of the most cunning charlatans are for-profit telemarketers that call on behalf of sick kids, veterans, homeless animals, and other softhearted issues. You may think it's a simple way to donate. It is—for the telemarketers. People waste billions of dollars on inefficient, poorly run, or downright fraudulent charities because they do not bother to research where their money is going. For instance, in 2013, for-profit telemarketers persuaded New Yorkers to donate $302 million to charity. More than half of that money, $156 million, went straight to the telemarketing firms, rather than the charities for which they were collecting. It was spent on salaries, offices, three-martini lunches, business junkets, and other overhead costs, leaving far less than donors might expect for the people they wished to help.

People also can waste their good intentions by giving to organizations that have brand recognition without asking whether donations will accomplish their stated goals. For-profit telemarketing companies often keep 20 to 90 cents for each dollar, according to Charity Navigator.

The catch-22 here is that there are no laws governing how much of a donation a professional telemarketer can keep, so a targeted charity—no matter how supposedly well known they are—will probably see only pennies on the dollar from telemarketers They will pluck at your hearts strings, play on your sympathies, and pull dollars out of your bank account. The best advice is to tell them that you have already given and hang up.

CHECK OUT A CHARITY BEFORE WRITING A CHECK

A good public resource for investigating charities is the Foundation Center, a nonprofit organization whose stated mission is "to strengthen the social sector by advancing knowledge about philanthropy in the U.S. and around the world." Some charities publicize their ratings, such as The Nature Conservancy, whose annual report shows that 6 percent of their funds go to fund-raising and membership, 9 percent to administrative costs, and the rest to their stated goals.

"The typical charity spends 75 percent of its budget on programs, 10 percent on fundraising and 15 percent on administrative," says Sandra Miniutti of Charity Navigator. Charitable groups have bills to pay like everyone else, but "follow the cash," cautions Daniel Borochoff, president of the American Institute of Philanthropy, which runs CharityWatch.org.

If the charity's considerations and percentages seem too demanding, ask a representative for specific information, such as how many individuals were served or what were the major accomplishments during the past year.

Different types of charities have different resource and spending requirements. Let's take the American Cancer Society as an example. According to Charity Navigator, the ACS is probably one of the best-run major charities. For every dollar, 9.8 cents goes to administrative costs and 21.8 cents goes toward furthering their marketing efforts. On the other hand, nonprofit museums have above-average administration costs as compared to other types of charities because it is costly for them to maintain their facilities and collections. A charity should not be shy about explaining its costs, and it may be a legal responsibility to disclose their costs vs. amounts used for their causes.

If you really want a hands-on approach to your charitable donations look to organizations like Charity Navigator, an independent and well-regarded organization that judges charities and grades them in several categories, including how much money they rake in and how it is spent.

CharityWatch (https://www.charitywatch.org), founded 25 years ago as the American Institute of Philanthropy (AIP), is an independent, charity watchdog. They dive deep to let you know how efficiently a charity will use your donation to fund the programs you want to support, and they expose abuses.

Consumer Reports lists some of the best and worst bangs for your charity bucks based on various factors (0 to 4 stars), including how much of their money is spent on programs as opposed to fundraising or other purposes.

Another thing that charity watchdogs look at and that yanks people's chains is the salary of the CEO, officers, and staff. It's understandable that you want to invest in a cause, not an executive's pocket. The packages paid to some of the chief executives of big-time charities are shocking. Just because they run a "nonprofit" company does not mean that they are not compensated very sweetly. Those salaries can be checked out at Charity Navigator and Charity Watch.

GUIDEPOSTS FOR GIVING

Americans seem to be a selfless people. Whenever there is a disaster, there are contributions flowing in from everywhere in our country to help. But the fact of the matter is that although the United States is the 13th-wealthiest country by population, the Johns Hopkins Comparative Nonprofit Sector Project estimates that its charitable giving was 1.85 percent of the size of the economy in recent years.

One of the most interesting statistics related to charitable giving is about who is most likely to donate to charities. Ironically, the lowest and highest earners donate larger portions of their incomes relative to those of middle income. Perhaps this is because those with lower incomes are more sensitive to hardship and difficulties, and higher earners crave more social recognition from others, and have more discretionary income.

Americans with adjusted gross incomes between $100,000 and $200,000 contributed just under $3,400, or 4.2 percent of their discretionary income. Those who earned $50,000 to $100,000, by contrast, contributed about $2,000, or 6 percent of optional income, while people who earned more than $200,000 gave roughly $14,000, or just over 4 percent, according to the Chronicle of Philanthropy.

Who and How to Help

Our minds and morals are constantly being bombarded with pleas for aid to a myriad of causes and cures, pandering to our prejudices, vanities, concerns, and desires. It's a little like government pork-barrel spending in that everyone has a hand out for a handout. Many are the deserving poor who need our help. And some are even the deserving wealthy. Universities and other institutions often receive gifts for things like new techno-gadgets and grants for various studies that may offer large returns in technology and important discoveries in the future that will benefit everyone.

The fastest-growing sector of U.S. private philanthropy in recent years has been aid to poor people overseas. They get more help from private donors ($39 billion per year) than from official government aid ($31 billion). Poor countries need our help, but the target counts more than the thought. Charitable donations of used clothes caused the clothing industry across Africa to suffer a 40-percent decline in production and a 50-percent jump in unemployment from 1981 to 2000. Being able to change the world just by cleaning out your closet is a nice thought, but also a massive misunderstanding of what developing countries actually need or want.

There needs to be a long-term strategy as well. Bringing education to impoverished children in third world countries takes more than getting a school built. It includes how it will be maintained for the next several decades, where the teachers are going to come from, buying school supplies, and feeding the children if need be. Property taxes, upkeep, and general overhead should be part of the picture for charities like Habitat for Humanity, which supplies housing to families in need of decent, affordable housing.

"Awareness campaigns" are a popular method of raising funds, but they may be a waste of time. Wearing a pink ribbon or a T-shirt emblazoned with a logo, or carrying a banner in a parade can help draw the public's attention, but not as much help as people imagine. When Breast Cancer Awareness month rolls around, we think of pink shoes in the NFL, or hot chicks with boob-related t-shirts, and pink buckets of

KFC. But studies show that participation in awareness campaigns makes people less likely to give actual money. Doing a good PR deed can be an excuse to cut back on charitable donations because people feel that they have already given—at least of their time and calories.

Americans give roughly $358 billion per year. But only about 20 percent of that goes to health and human services causes. The rest goes to religion, higher education, and hospitals. A third of American charitable giving went to support religious institutions to pay for clergy salaries, the maintenance of structures, and administration. The next largest category was for higher education. So, choose wisely and give to causes that appeal to you and not the boss, neighbor, boy or girl friend, or buddy.

Can the Canned Food Drives

Canned food drives sound good but are basically an ill-advised solution to hunger, and it leads to giving people a chance to dump stuff they won't ever eat. The local charitable food pantry really doesn't want two-year-old steamed okra, flaked herring, or canned haggis. First, it has to be trucked to a warehouse then sorted and all the expired cans have to be tossed out. Half the foods donated do not meet nutritional standards. Also, canned foods and many boxed processed foods are notoriously high in sugar and salt, which is not good for older folks and those with diabetes.

The best way to aid a food bank is to donate money that allows them to buy perishables like fresh meats, veggies, and bread at a wholesale level and from farmers. The can of beans that might cost a buck is not as efficient as buying in bulk.

Ambushed

You're in line and have a credit or debit card out waiting for the cash register reader. Suddenly, the machine or the checker asks if you would like to make a donation, and you're caught off guard. You don't want to look like a miser, especially during the holidays. It requires a snap decision, which doesn't give you any time to properly evaluate the charity or how the funds are going to be used. You feel hijacked. Welcome to "charity checkout ambush."

More and more people caught by checkout charity are getting angry and feeling like they are being held up. It's a brilliant, cost-effective, no-brainer for charities. And during a shopping expedition, it's possible to get hit several times in one afternoon.

A recent CBS poll showed more than 82 percent of respondents complained that checkout charity feels like being bushwacked. And few people are willing to save a grocery receipt for a tax deductible dollar. An analysis of more than 60 of the most prominent fund-raising campaigns discovered that checkout charities brought in $358 million in 2012, according to consulting firm Cause Marketing Forum.

Then there is the yearly tradition of the Girl Scout Cookie drive, a pretty little program worth about $800 million annually. About 15 percent goes to a GS troop, around 5 percent goes to a regional organization, and 20 percent goes to the cost of goods. And the no-return policy is so strict that girls and their families often end up literally eating the cost of their leftover inventory.

The remaining 60 percent is an enviable profit margin, especially since cookies are sold exclusively by a cute, diminutive, and free sales force with an army of adults for backup. It's a brilliant business model made possible through child labor, restrictive return policies, protective parents, and soft-hearted adults. All in all, it's a pretty sweet-tasting deal.

Getting Worked at Work

Someone at work has a pet project or volunteers to head up a fund-raising drive. There are informational meetings, pep rallies, or desk-to-desk solicitations. It's hard not to succumb in the workplace. For some people, it's a convenient way to give, but for others it's a time to lower their head in their cubicle and avoid the break room and emails.

There are many ways to make charitable gifts work at work, such as participating in a sport to raise money for a nonprofit, having donations matched by your company, or simply collecting donations to benefit victims of a local disaster. But "workplace giving" should refer to a program that usually happens once per year, includes a payroll-deduction option, and allows you to make some choices of your own.

Think Long Term

Determining how education could be brought to impoverished children in third world countries is more than getting a school built. It includes how it is going to be maintained for the next several decades, where the teachers are going to come from, buying school supplies, and how to feed the children if need be, all should be part of the overall picture. Property taxes, upkeep, and general overhead should be part of the picture for charities like Habitat for the Humanity, which supplies housing to families in need of decent, affordable housing.

Attaching Strings to Gifts

We've all experienced getting "conditional" gifts that sometimes impose obligations from political or religious organizations. It can make a difference if the conditions imposed conform to an opinionated party or a belief.

The reasons most people give for objecting to conditional gifts are:

- Conditions interfere with the autonomy of the recipient and the donator.
- Conditions interfere with self-determination or the rights of sovereign states.
- Conditions are contrary to human rights.
- Conditions are manipulative.

THE PROBLEM OF PANHANDLING

When looking into the alms cup of a homeless person, people are bound to feel many things—empathy to enmity, hubris to humility. The urge to give may be tempered by the apprehension that your money might be used for the things that keep people addicted and on the streets—alcohol and drugs.

We're all familiar with the street signs seen in many cities, "Give to charities that help the poor. Handouts will only encourage destructive behavior." And it may be true. A report from the Department of Housing and Urban Development (HUD) found that six out of ten homeless people interviewed admitted problems with substance abuse. The true stats are probably higher. Living on the streets is tough. But that's only part of the problem. The study also reported that 94 percent of money panhandled was used to purchase food. It also found 60 percent of panhandlers made $25 a day or less, breaking the myth that beggars make large amounts of money.

City social service researchers spoke with 400 San Franciscans that had given money to panhandlers. They found that the largest group of people who chose to give were young working-class residents. Empathy was a main driver; three in five said they gave because they or a family member might be in need someday.

Practical Panhandling

It's what's called "giving a hand up, not a hand out." It's the theory that it's best not to give money directly to street people. If a person truly wants to eat or find clothing, food, or shelter, then almost all cities offer help. You can give tangible, usable items such as gift certificates to McDonalds, clean socks, canned food, fresh fruit, small jars of jam and peanut butter, single-serving drinks, or bottled water. Travel-size toiletries—shampoo, lotion, deodorant, razors, baby wipes, toothpaste, toothbrushes, tissue packets, are also a good idea.

Let the experts do what they do best; they have knowledge and experience in social work. Being immersed in the world of helping others, they continually strategize about how to be effective. Experts say that homeless people are seldom dangerous or to be feared. They're usually just downtrodden people looking for help. But if you do stop and give, just like with any unpredictable situation, care needs to be taken when interacting with strangers, particularly if there's the possibility that drugs and/or alcohol are in the mix.

Reasons it's okay to give to homeless people:

- A little money may not mean much to you, but it could mean a lot to someone else.
- It encourages kindness and creates good vibes.
- You believe that giving is kind, no matter the outcome.
- You don't know a homeless person's story.
- It could be you, or someone you know someday.

- When someone does a good deed for you, pay it forward.

Panhandling and the Law

Business owners are the most likely demographic to contact the police when panhandlers are outside of a business, as they claim it costs them customers. But enforcing panhandling laws has become expensive for many cities with limited resources. And enforcement doesn't solve the problem because the homeless are still around and won't go away.

Panhandling costs

In Colorado, the choice has largely been to criminalize homelessness—a choice that comes with a big price tag—more than $750,000 per year, and six other cities have spent more than $5 million enforcing 14 anti-homeless ordinances over the last five years through policing, court, and incarceration costs.

The legal battle of begging

Although panhandling is protected by the First Amendment in the United States, many communities are actively trying to outlaw it, and a fast-growing body of legal action testifies to the fact that the criminalization has become a kind of legal battleground. It's creating clashes between social hygiene and freedom of expression; middle-class discomfort and underclass economic need; commercial interests of downtown business owners and beggars' rights to plead for subsistence.

As a way to alleviate concerns and regulate the growing numbers of panhandlers, some municipalities require a license to panhandle. All you need is a valid ID and to follow some rules and regulations.

No one really knows whether a gift or a dollar here and there really helps. But like the little boy who was asked why he rescued a starfish from low tide when there were so many others he said, "Why don't you ask the one I saved."

SOURCES

"12 Common Criticisms of Philanthropy—and Some Answers." Philosophy Roundtable, http://www.philanthropyroundtable.org/topic/excellence_in_philanthropy/12_common_criticisms.

Dickens, Charles. "A Christmas Carol." London: Chapman & Hall, 1843, https://en.wikipedia.org/wiki/A_Christmas_Carol.

Alexander, Dan. "How Donald Trump Shifted Kids Cancer Charity Money into His Business." *Forbes*, June 6, 2017, https://www.forbes.com/sites/danalexander/2017/06/06/how-donald-trump-shifted-kids-cancer-charity-money-into-his-business/.

Altucher, James. "10 Unusual Ways to Release Oxytocin into Your Life." https://jamesaltucher.com/2012/08/10-ways-release-oxytocin-into-life/.

Andresen, Katya. "Science of Giving 8: The Identifiable Victim Effect—and Its Limits." Non Profit Marketing (blog), http://www.nonprofitmarketingblog.com/site/science_of_giving_8_the_identifiable_victim_effect_-_and_its_limits/.

Arvind, Y. "A Starfish and the Ocean." Heart to Heart, June 2018, http://media.radiosai.org/journals/Vol_06/01JUN08/09-healingtouch.htm.

Brady, Edith. "Business Owners Face Challenges with Some Homeless." *Pantagraph*, May 19, 2016, http://www.pantagraph.com/news/local/business-owners-face-challenges-with-some-homeless/article_f947de2a-3599-50be-8b79-b07700142ab6.html.

Becker, Ernest. *The Denial of Death*. New York: Free Press, 1997.

Bodhi, Bhikku. "Dana: The Practice of Giving." Access to Insight, 1995, https://accesstoinsight.org/lib/authors/various/wheel367.html.

Burak, Jacob. "Is Philanthropy Driven by the Human Desire to Cheat Death?" Big Think, November 20, 2017, http://bigthink.com/aeon-ideas/is-philanthropy-driven-by-the-human-desire-to-cheat-death.

Carlson, Joan. "Wiley Panhandler Oscar Finds a Soft Touch." *Kitasap Sun*, October 19, 2003, https://web.kitsapsun.com/archive/2003/10-19/286825_wiley_panhandler_oscar_finds_a_.html.

CharityC. "What Food Banks Need Most (and What They Can't Use)." Foodlets (blog), November 18, 2017, http://foodlets.com/2014/11/18/what-food-banks-need-most-and-what-they-get-too-much-of/.

Covert, Bryce. "States That Criminalize Homelessness End up Paying the Price." Think Progress, February 24, 2016, https://thinkprogress.org/states-that-criminalize-homelessness-end-up-paying-the-price-319234ef2be8/.

Depew, Bradley. "Learn about Millennials and Charity." Balance Small Business, March 5, 2018, https://www.thebalancesmb.com/how-millennials-have-changed-charitable-giving-2501900.

Dewey. "Beggars Really Do Need Your Help." Daily Revolution, November 6, 2013, http://dailyrevolution.com/2013/11/06/beggars-really-do-need-your-help/.

"Executive Salaries in Charities." Snopes, https://www.snopes.com/fact-check/executive-salaries-charities/.

Ginns, Bernard. "Is Donating to Large Charities a Waste of Money? *Spectator*, September 2016, https://blogs.spectator.co.uk/2016/09/charities-executive-pay/.

"Guide to Giving in the Workplace." Charity Navigator, April 17, 2006, https://www.charitynavigator.org/index.cfm?bay=content.view&cpid=159.

Hall, Holly. "Americans Rank 13th in Charitable Giving among Countries around the World." *The Chronicle of Philanthropy*, December 3, 2013, https://www.philanthropy.com/article/Americans-Rank-13th-in/153965.

Hill, Mark. "5 Popular Forms of Charity That Aren't Helping." *Cracked*, July 1, 2012, http://www.cracked.com/article_19899_5-popular-forms-charity-that-arent-helping.html.

James Baraz, Shoshana Alexander. "The Helper's High." Big Ideas, February 1, 2010, https://greatergood.berkeley.edu/article/item/the_helpers_high.

Johnson, Holly. "5 Reasons It's Okay to Give Money to Homeless People." Club Thrifty, April 29, 2015, https://clubthrifty.com/5-reasons-to-give-money-to-homeless-people/.

Kane, Colleen. "Where Are Your Charity Dollars Going?" CNBC News, December 9, 2010, https://www.cnbc.com/id/40592354.

Keyes, Scott. "Everything You Think You Know about Panhandlers Is Wrong." Think Progress, October 30, 2013, https://thinkprogress.org/everything-you-think-you-know-about-panhandlers-is-wrong-36b41487730d/.

Knight, Heather. "The City's Panhandlers Tell Their Own Stories." *SF Gate*, October 27, 2013, https://www.sfgate.com/bayarea/article/The-city-s-panhandlers-tell-their-own-stories-4929388.php.

Kurtzleben, Danielle. "How Girl Scout Cookie Sales Stack Up against Big Brands." *U.S. News and World Report*, January 16, 2013, https://www.usnews.com/news/articles/2013/01/16/how-girl-scout-cookie-sales-stack-up-against-big-brands.

Libresco, Leah. "Charities Don't Need to Break Laws to Waste Your Money." *Five Thirty Eight*, May 21, 2015, https://fivethirtyeight.com/features/charities-dont-need-to-break-laws-to-waste-your-money/.

Linden, David J. "This Is Your Brain on Charitable Giving." *Psychology Today*, August 31, 2011, https://www.psychologytoday.com/us/blog/the-compass-pleasure/201108/is-your-brain-charitable-giving.

"Major Achievements of American Philanthropy, Religion." Philanthropy Roundtable, http://www.philanthropyroundtable.org/almanac/religion.

"Making Your Charitable Gifts Last." Wilmington Trust, https://library.wilmingtontrust.com/wealth-planning/making-your-charitable-gifts-last.

"Middle-Income Earners Probably Won't Be Paying as Much Tax as the Government Expects." *The Conversation*, October 16, 2017, http://theconversation.com/middle-income-earners-probably-wont-be-paying-as-much-tax-as-the-government-expects-85672.

Minoguchi, Hajime. "Why Canned Food Drives Should Not Be Promoted." *The Leaf*, December 11, 2015, https://shsleaf.org/23653/opinion/why-canned-food-drives-should-not-be-promoted/.

National Coalition for the Homeless and the National Law Center on Homelessness & Poverty, *A Dream Denied: The Criminalization of Homelessness in U.S. Cities*. www.nationalhomeless.org/publications/crimreport/report.pdf.

"Not for Profit Organizations, A Legal Guide," http://www.legal-info-legale.nb.ca/en/publications/consumer_law_and_non_profit/non_profits/not_for_profit_organizations_legal_guide.pdf.

Park, James. "Checkout Charity: Get Ready for the Cash Register Ambush." *Finacial Post*, December 3, 2014, http://business.financialpost.com/personal-finance/checkout-charity-get-ready-for-the-cash-register-ambush.

Praats, Michael. "Panhandling Is Protected by the First Amendment, Aggression Is Not." *Port City Daily*, January 28, 2018, https://portcitydaily.com/local-news/2018/01/28/wilmingtons-panhandlers-are-protected-by-the-first-amendment-until-theyre-not-suggestion-nws/.

Radde, Tom. "Giving USA: Americans Donated an Estimated $358.38 Billion to Charity in 2014; Highest Total in Report's 60-Year History." Giving USA, June 29, 2015, https://givingusa.org/giving-usa-2015-press-release-giving-usa-americans-donated-an-estimated-358-38-billion-to-charity-in-2014-highest-total-in-reports-60-year-history/.

Reeser, Devon. "Charity Navigator vs. CharityWatch: What Your Nonprofit Needs to Know." Fundraising IP, http://www.fundraisingip.com/fundraising/charity-navigator-vs-charitywatch-nonprofit-needs-know/.

Ruis, Rebecca. "4 Cancer Charities Are Accused of Fraud." *New York Times*, May 19, 2015, https://www.nytimes.com/2015/05/20/business/4-cancer-charities-accused-in-ftc-fraud-case.html.

Scott, Michael S. "The Problem of Panhandling." Center for Problem Oriented Policing, http://www.popcenter.org/problems/panhandling/print/.

Singer Peter. "Charity: The Psychology of Giving." *Newsweek*, February 7, 2009, http://www.newsweek.com/charity-psychology-giving-82369.

"Statistics." Philanthropy Roundtable, http://www.philanthropyroundtable.org/almanac/statistic.

Taylor, Chris. "Marketing Experts: Stores Ambush' Customers to Donate Money at Cash Register." *Huffington Post*, December 6, 2017, https://www.huffingtonpost.com/2014/11/25/charity-cash-register_n_6213834.html.

Taylor, James. "The Great Checkout Counter Charity Grab." *Fiscal Times*, November 24, 2014, http://www.thefiscaltimes.com/2014/11/24/Great-Checkout-Counter-Charity-Grab.

"The Right Amount to Donate to Charity." *Money*, September 14, 2102, http://time.com/money/2792145/the-right-amount-to-donate-to-charity/.

"The Problem with Panhandling." WHOI TV, November 9, 2017, https://www.whio.com/news/the-problem-with-panhandling/JEp7etezP7LsgA5UXFwPOO/.

"The Power of One: The Psychology of Charity." Association for Psychological Science, https://www.psychologicalscience.org/news/were-only-human/the-power-of-one-the-psychology-of-charity.html.

Thompson, Derek. "Should You Give Money to Homeless People?" *Atlantic Monthly*, March 22, 2011, https://www.theatlantic.com/business/archive/2011/03/should-you-give-money-to-homeless-people/72820/.

Tocqueville, Alexis de. *Democracy in America*. Cited in Wikipedia, https://en.wikipedia.org/wiki/Democracy_in_America.

Weston, Liz. "6 Mistakes People Make When Giving to Charity." *Money*, November 25, 2015, http://time.com/money/4127069/charity-giving-mistakes/.

Wiltz, Teresa. "Anti-Panhandling Laws Spread, Face Legal Challenges." *State Line*, November 12, 2015, http://www.pewtrusts.org/en/research-and-analysis/blogs/stateline/2015/11/12/anti-panhandling-laws-spread-face-legal-challenges.

Zinsmeister, Karl. "Giving Helps the Giver Too." *Huffington Post*, March 5, 2017, https://www.huffingtonpost.com/karl-zinsmeister/giving-helps-the-giver-to_b_9387064.html.

9

PAPER WASTE CHASE

> Great ideas are started on paper. The world is educated on paper. Businesses are founded on paper. Love is professed on paper. Important news is spread on paper. Justice is rendered on paper. Rights are guaranteed on paper. Freedoms are declared on paper.
> —*Printed on the cover of a ream of paper*

The history of paper dates back almost 6,000 years to the Egyptians, who crafted Papyrus, the root of the English word for paper, made by weaving reeds or other fibrous plants together and pounding them into a flat sheet. The ever-ingenious Chinese first invented what is considered to be the first modern-day paper sheets out of mulberry bark, hemp, rags, and water. For hundreds of years afterward, paper was a rare and precious commodity and revealing the secret of papermaking was punished by death. Now, it overfills our planet and threatens to bury us.

It has been estimated there are 20,000 identifiable uses for paper in the world today—everything from creating literature to printing currency to smoking various substances in it. In one of his *Tonight Show* monologues, Johnny Carson made a joke that there was a shortage of toilet paper, which created panic buying and made the front page of the *New York Times*.

People thought that the electronic age would usher in a paperless era. This could not be further from the truth. A wag once quipped that the electronic age's challenge to lessen the use of paper was about as plausible as a paperless bathroom. We use almost 26 percent more paper than we did 20 years ago and paper waste amounts to almost 30 percent of our solid waste municipal landfill. We are the largest producer and consumer of paper, and the largest producer of paper waste.

Facts and figures:

Nearly 4 billion trees or 38 percent of the total trees cut around the world are used in paper industries on every continent, and about 28 percent of all wood cut in the United States is used for papermaking. The demand for pulp and paper products will more than double by 2060. Understanding the facts about paper consumption is the first step to a paperless, or at least a less paper-inundated environment.

- 93 percent of paper comes from trees.
- 50 percent of the waste of businesses is composed of paper, and it is estimated that 40 percent of the waste in the United States is paper.
- Paper in the United States represents one of the biggest components of solid waste in landfills—250 million tons, or 27 percent of landfill solid waste, and accounts for 25 percent of landfill waste and 33 percent of municipal waste.
- The waste created by paper could build a 20-foot-high wall all along the Mexican border, with paper to spare (President Trump take note).
- Close to 40 percent of the world's industrial logging goes into making paper, and this is expected to reach 50 percent in the near future.
- In the last 20 years, the usage of paper products has increased from 92 million tons to 208 million, a growth of 126 percent.
- Americans still consume and waste more paper per capita, more than 749 pounds of paper every year.
- On average, an adult uses over 20,000 sheets of toilet paper a year and more than 3,000 tons of disposable paper towel waste.
- In order to produce one ton of paper towels, 17 trees are cut down, and 20,000 gallons of water are used in the process, more than any other industry. It takes an average of five liters of water to produce one piece of paper.
- A hundred million trees are ground up each year to produce junk mail, and American's receive almost 4.5 million tons of junk mail per year, about 44 percent of which is never opened.

- It takes twice the energy to produce paper than to produce plastic bags.
- Recycling a ton of paper saves around 682.5 gallons of oil, 26,500 liters of water, and 17 trees.
- In the United States, companies spend more than $120 billion a year on printed forms, most of which outdate themselves within three months time and are never used.
- We recycle about 38 percent of all paper.
- Paper in the average business grows by 22 percent a year, meaning paper waste will double every 3.3 years.

PUSHING PAPER IN THE OFFICE

Each year, the world produces more than 300 million tons of paper. According to the Environmental Protection Agency (EPA), printing and writing papers typically found in a school or office environment such as copier paper, computer printouts and notepads, comprise the largest category of paper product consumption. This is an alarming figure considering today's advanced computer technology was specifically designed to help offices reduce paper use.

The average office worker generates about two pounds worth of mixed paper products every day and uses about 10,000 sheets of paper each year, 45 percent of the paper printed in offices ends up trashed by the end of the day, and amounts to over a trillion sheets of paper per year. Thirty percent of print jobs are never even picked up from the printer, and 70 percent of the total waste in offices is made up of paper. That amounts to more than $120 billion a year spent by companies on paper—much of it wasted

PAPER CONSUMPTION—THE ENVIRONMENT

Although recycling is important, it doesn't keep prodigious piles of paper from the landfills. According to Zero Waste New Zealand Trust, a nonprofit organization, making paper from recycled content creates 74 percent less air pollution and 35 percent less water pollution. Paper cannot be recycled indefinitely, and seven to ten times is about the maximum before the fibers break down, so there will always be a need for virgin wood pulp.

Paper pollution is a serious problem. The life cycle of paper is damaging to the environment from beginning to end. It starts off (partially) with trees being cut down, stopping their ability to absorb CO_2, and sometimes ends its life by being burned and emitting greenhouse gases into the atmosphere—however, the CO_2 problem is a wash when considering the amount it removed from the atmosphere when a tree is growing.

It is estimated that by 2020, paper mills will be producing a monstrous 500,000,000 tons of paper and paper products each year.

A BAD WRAP FOR FAST FOOD

Clean Water Action (CWA) decided to learn more about the sources of trash that end up in the San Francisco Bay Area. The nonprofit decided to collect samples of street litter from the Bay Area cities of Oakland, Richmond, San Jose, and South San Francisco. CWA found that the biggest source, 49 percent, of litter is from fast-food outlets. The five most significant sources were McDonalds, Burger King, 7-Eleven, Starbucks, and Wendy's. Starbucks is the largest user of coffee cups in the world—2.946 billion cups at their stores, or an average of 8,070,428 per day.

It lists in its Global Responsibility Report 2010 the goal of making 100 percent of its cups reusable or recyclable in its more than 17,000 retail sites worldwide. Starbucks cups are still not recyclable at this time. However, the company has teamed up with Closed Loop Partners, which invests in sustainable consumer goods, to develop a recyclable and compostable cup. The companies have committed $10 million to this venture, and their goal is to produce a cup that can be manufactured, used, fully recycled, and then made back into cups. But it hasn't happened just yet.

McDonald's appears to be backsliding on its long-overdue assurance to use more environmentally friendly packaging, having reintroduced foam cups that aren't biodegradable and that are tough to recycle. However, almost twenty years ago, it ended foam "clamshell" packaging, used for hamburgers and sandwiches, and switched to paper wrapping—which ironically has caused more problems. Among McDonald's fast-food peers, only Dunkin' Donuts and Chick-fil-A haven't made the switch to paper cups.

Next time you drop in for a fast food fill-up think about what else besides a Happy Meal you might be putting in your body. Most of the time, when you order fast food, you kinda know what you're getting as far as sodium, sugar, fat, and cholesterol. But what you don't know is what's in the packaging wrapped around food that could give your health a bad rap, according to a report by the Silent Spring Institute and published in the journal *Environmental Science & Technology Letters.*

The menace is called perfluorinated chemicals (PFC), fluorines used and found in nearly half of the food contact papers and 20 percent of the paperboard containers you handle. They're used because they repel grease in food wrapping. PFCs are a family of nonstick, waterproof, stain-resistant chemicals that DuPont and other companies have used since the 1950s. Teflon and 3M's Scotchgard are the best-known PFC-based products.

More than a decade since the health hazards of PFCs became known, many fast food chains still use food wrappers, bags, and boxes coated with the grease-resistant compounds, according to nationwide tests reported in a peer-reviewed study by the *Journal of Exposure Science and Environmental Epidemiology* .

Food contact paper was the worst, with 46 percent of all samples testing positive for fluorine. Food contact paperboard (like pizza boxes and to-go packages) were next, at 20 percent, followed by beverage containers at 16 percent. Non-food contact paper and paper cups all tested negative for fluorine.

The report found that the most studied of these chemicals, perfluorooctanesulfonates (PFOSs) and perfluorooctanoates (PFOAs), have been linked to kidney and testicular cancer, elevated cholesterol, decreased fertility, thyroid problems, changes in hormone functioning (affecting growth and gender assignment), and decreased immune response in children, who are especially at risk for health effects because their developing bodies are more vulnerable to toxic chemicals.

The research also has shown that chemicals in food packaging migrates when discarded in landfills and can contribute to elevated levels of toxins in the environment. To make matters worse, the chemicals resist degrading and will remain in the environment for quite a while.

Some cities found that the biggest source of litter is fast-food packaging that lands in dumps or waterways and can leach chemicals into groundwater.

Short of asking that your next fast-food meal be served sans fast-food wrapping, in a paper cup, or a noncontact paper bag, there isn't a whole lot you can do, short of taking the food out of the packaging sooner rather than later.

It should be noted that unless you eat fast foods at an alarming rate and the paper that goes along with it, although neither is healthy, you would have to consume a lot of both before you might suffer serious health reactions.

The FDA has approved 20 next-generation PFCs specifically for coating paper and paperboard used to serve food. But these chemicals have not been adequately tested for safety, and trade secrecy laws mean that this information is not disclosed to the public. Retired EPA Toxicologist and Senior Risk Assessor Deborah Rice said, "The newer paper has the same constellation of health effects. There's no way you can call this a safe substitute." Have a happy meal.

PAPER POISONS

Unfortunately, the papermaking process is not a clean one. According to the EPA, pulp and paper mills are among the worst polluters of air, water, and land of any industry in the country. The papermaking industry uses toxic chemicals like chlorine and chlorine compounds, mercury, and toxic organic halides (a toxic compound that contains fluorine, chlorine, bromine, or iodine) and produces waste of total suspended solids (mostly decaying paint matter), boiler ash, and effluent sludge (organic and chemical residue). Other pollutants include toxins such as bleach, carbon monoxide, ammonia, nitrogen oxide, nitrates, methanol, benzene, dioxins, and chloroform. None of these chemicals is good for living things.

Environmental priority is trickling down. Staples, for example, announced that it would reduce sales of paper made of wood from endangered forests by 50 percent, and it is committed to move the majority of their paper products to the Forest Stewardship Council (FSC). The FSC supports environmentally appropriate, socially beneficial, and economically viable management of the world's forests. Office Depot announced that the company's top-selling 30 percent recycled content paper combines responsible FSC forestry and recycled content.

The Environmental Paper Assessment Tool is a new website to help buyers and sellers of paper products evaluate and select "environmentally preferable" paper, which takes into consideration toxicology as well as recyclability and other factors.

WASTED WOOD AND DEFORESTATION

In the past, papermaking preferred virgin forests and old-growth trees for pulp, which take hundreds of years to grow and are virtually nonrenewable. Due to the needs of construction over the last 100 years, much of our virgin forests were clear-cut. These days, the paper industry has moved toward renewable "plantation" forests for pulp paper.

Trees are a renewable resource, which means that once one is cut down another can be planted in its place and much of the wood used by paper companies in the United States comes from privately owned tree farms, where forests are planted, groomed, and thinned for harvest in 20- to 35-year cycles. But increasing demand for wood products is a driver of tropical deforestation. In fact, there are more trees in the United States than there were a hundred years ago. But regrowth is far different than virgin forests. (See more in chapter 4 on Losing Land.)

Prodigal Paper Waste

There are alternatives, such as agricultural waste as a stand-in for wood. Agri-pulp is made of wheat, oat, barley, and other crop stalks left over after harvesting. Combined with recycled paper and other fillers, some papermakers are finding that agri-pulp paper makes fine stationery. Other fibers include cotton, banana, tobacco, citrus, coffee bean, bamboo, bagasse (the dry pulpy residue of sugar cane and sorghum), and recycled fabrics. Hemp, on which the Declaration of Independence was written, is a wood substitute that has a rich history in papermaking. And would you believe that some paper is being made out of elephant dung?

About 77 percent of all papermakers in the U.S. substitute some recycled paper for virgin wood in the pulp-making stage, and they use only recycled waste as their primary source of raw material.

The pulp and paper industry is one of the heaviest users of water within the country according to industry experts. Approximately 85 percent of the water used in the industry is used as processed water, resulting in relatively large quantities of contaminated water and necessitating the use of on-site wastewater treatment solutions. The EPA is assigned the oversight in the papermaking industry.

Reducing Paper Waste and Pollution

A small effort on all our parts will be a valuable contribution in the resolution of our paper pollution problem. When people join together and demand forest conservation, recycling, and a cleaner production, businesses and governments will listen.

Be a conscientious consumer and buy 100 percent recycled paper products. Make sure that products made from forest fiber bears a seal from a credible forestry certification system, like the FSC, which helps the move toward zero deforestation.

Uses both sides of paper, this small effort reduces paper waste by 50 percent. If you already have a scanned copy of a file, don't print it unless it's really needed. Use email instead of paper when communicating with clients and customers.

Cut down on the use of paper cups and disposable paper plates by keeping reusable items in the office pantry. Buy products with the least paper packaging. Take advantage of the latest technologies like tablets, computers, and smartphones to keep your files and notes in order.

Ask local and national governments for mandatory source reduction plans and programs. Extended Producer Responsibility (EPR) is a resource management tool whereby producers assume responsibility for the end-of-life management of their used products. This can include collection, sorting, and treating products for their recovery and recycling. Ask for minimum recycled content packaging requirements and landfill bans on recyclable packaging.

Remember that the best use of paper is to use less paper.

SOURCES

"Are the People of Earth Consuming Trees Faster Than They Can Be Replaced, or Are We Not?" *Quora*, https://www.quora.com/Are-the-people-of-earth-consuming-trees-faster-than-they-can-be-replaced-or-are-we-not.

Andrews, Dave, and Bill Walker. "Many Fast Food Wrappers Still Coated in PFCs, Kin to Carcinogenic Teflon Chemical." Environmental Working Group, February 1, 2017, https://www.ewg.org/research/many-fast-food-wrappers-still-coated-pfcs-kin-carcinogenic-teflon-chemical#.Wt_FSjNlDcs.

Basbanes, Nicholas A. "10 Most Bizarre Uses of Paper in History." *Huffington Post*, December 21, 2013, https://www.huffingtonpost.com/nicholas-a-basbanes/post_5940_b_4136654.html.

Bodkin, Harry. "Warning: Fast Food Packaging and Grease-Proof Paper Contain Potentially Harmful Chemicals." *Telegraph*, February 1, 2017, https://www.telegraph.co.uk/science/2017/02/01/warning-fast-food-packaging-grease-proof-paper-contain-potentially/.

Boyle, Lisa Kaas. "Chick-fil-A Is Full of It: Styrene Is Not Green." Planet Experts, January 14, 2016, http://www.planetexperts.com/chick-fil-a-is-full-of-it-styrene-is-not-green/.

Cheeseman, Gina Marie. "Fast Food Garbage Makes up 50% of Street (and Pacific Gyre) Litter." *Triple Pundit*, June 27, 2011, https://www.triplepundit.com/2011/06/fast-food-big-source-trash-pollution/.

"Cons of Hemp?" Marijuana, March 21, 2005, https://www.marijuana.com/community/threads/cons-of-hemp.161137/.

Coupal, Joseph. "The Environmental Impact of Paper Manufacturing." Atlantic Poly, Inc., December 20, 2013, http://atlanticpoly.com/Atlantic_Poly_Blog/the-environmental-impact-of-paper-manufacturing.

"Does Breathing Contribute to CO_2 Buildup in the Atmosphere?" *Skeptical Science*, https://skepticalscience.com/breathing-co2-carbon-dioxide.htm.

Digman, Larry. "Starbucks: Cup Recycling Logistics Harder Than They Seem." *ZDNet*, April 19, 2010, https://www.zdnet.com/article/starbucks-cup-recycling-logistics-harder-than-they-seem/.

Dolmage, Ami. "Shut the Front Door! If We Ate Less Meat, This Is What Would Happen to the Planet." *One Green Planet*, July 31, 2016, http://www.onegreenplanet.org/animalsandnature/eat-for-the-planet-meat-and-the-environment/.

"Environmental Impact of Paper." Wikipedia, https://wikivisually.com/wiki/Environmental_impact_of_paper.

Eves, Cris. "Timber and Pulp Companies Selected." *Spott*, June 2017, https://www.spott.org/news/timber-pulp-paper-companies-selected/.

Fredericks, Sean. "Earth Day Is Almost Here. How Does Your Paper Waste Measure Up?" Legal Shred, April 13, 2016, http://legalshred.com/earth-day-almost-paper-waste-measure-.

"How Many Trees Are Cut Down a Day for Paper?" *Quora*, https://www.quora.com/How-many-trees-are-cut-down-a-day-for-paper.

"If We Are Wasting One Paper, Does It Mean That We Are Wasting 100 Trees?" *Quora*, https://www.quora.com/If-we-are-wasting-one-paper-does-it-mean-that-we-are-wasting-100-trees.

LeBlanc, Rick. "Paper Recycling Facts, Figures and Information Sources." The Balance Small Business, February 6, 2017, https://www.thebalancesmb.com/paper-recycling-facts-figures-and-information-sources-2877868.

"Leading Coffee House Chains Ranked by Number of Stores Worldwide in 2015." *Statista*, https://www.statista.com/statistics/272900/coffee-house-chains-ranked-by-number-of-stores-worldwide/.

Martin, Sam. "Paper Chase." Geology Webinar, September 10, 2011, http://www.ecology.com/2011/09/10/paper-chase/.

Matthews, Lyndsey. "McDonald's Is Changing Its Packaging." *Delish*, January 18, 2018, https://www.delish.com/food-news/news/a57713/mcdonalds-eco-friendly-packaging/.

"Municipal Solid Waste." U.S. Environmental Protection Agency, https://archive.epa.gov/epawaste/nonhaz/municipal/web/html/.

"Office Depot Announces Its Top-Selling Recycled Copy Paper Is Now Forest Stewardship Council (FSC) Certified." Office Depot, Office Max, January 19, 2010, http://news.officedepot.com/press-release/products-and-services-news/office-depot-announces-its-top-selling-recycled-copy-paper-.

"Paper Comes from Trees . . ." The World Counts, May 13, 2014, http://www.theworldcounts.com/stories/Paper-Waste-Facts.

"Paper Industry—Statistics & Facts." *Statista*, https://www.statista.com/topics/1701/paper-industry/.

"Statistics." Paper Recycles, http://www.paperrecycles.org/statistics.

O'Mara, Morgan. "How Much Paper Is Used in One Day?" *Record Nations*, February 10, 2016, https://www.recordnations.com/2016/02/how-much-paper-is-used-in-one-day/.

"Paper Waste Is Ruining the Environment." *Mr. Rooter* (blog), March 2, 2017, https://www.mrrooter.com/about/blog/2017/march/paper-waste-is-ruining-the-environment/.

"Recycling." *Inc.*, https://www.inc.com/encyclopedia/recycling.html.

"Recycling." Encyclopedia.com, https://www.encyclopedia.com/science-and-technology/biology-and-genetics/environmental-studies/recycling.

Reed, Robert."McDonald's Foam Cups Running Over with Environmental Controversy." *Chicago Tribune*, July 7, 2017, http://www.chicagotribune.com/business/columnists/ct-mcdonalds-cup-troubles-robert-reed-0709-biz-20170706-column.html.

Rodriguez, Germania. "Dunkin' Donuts to Ditch Foam Cups by 2020 Removing One Billion Polystyrene Containers from the World's Waste but It Will Keep It in Its Logo." *Daily Mail*, February 10, 2018, http://www.dailymail.co.uk/news/article-5375895/Dunkin-Donuts-ditch-foam-cups-2020.html.

Seaver, Robert E. "Fluorine Compounds." Science.gov, https://www.science.gov/topicpages/f/fluorine+compounds.html.

"Space Shuttle Glider, *scribd/NASA*, https://www.scribd.com/document/35955182/Paper-Model-Space-Shuttle-Glider.

"Sustainable Products." *Staples*, https://www.staples.com/sbd/cre/marketing/about_us/sustainable-products.html.

"The Great Toilet Paper Shortage." *Useless Information*, http://uselessinformation.org/toilet_paper/index.html.

"The Top 10 Countries That Produce the Most Waste." IILyear4, https://sites.google.com/site/iilyear4/top-10-countries-that-produce-the-most-waste.

Tinker, Ben. "Report Finds Chemicals in One-Third of Fast Food Packaging." *CNN News*, February 1, 2017, https://www.cnn.com/2017/02/01/health/fast-food-packaging-chemicals-pfas-study/index.html.

"Trees Speak to Us. Let's Speak for Them." *Care2 Petitions*, https://www.thepetitionsite.com/937/610/895/trees-speak-to-us.-lets-speak-for-them.

Wilson, Mark. "The World's Largest Starbucks Is the Willy Wonka Factory of Coffee." *Co.Design*, December 4, 2014, https://www.fastcodesign.com/3039419/the-worlds-largest-starbucks-is-the-willy-wonka-factory-of-coffee.

10

PLASTICS

A Blessing and a Curse

"I want to say one word to you. Just one word."
"Yes sir."
"Are you listening?"
"Yes I am."
"Plastics."
—*The Graduate*

There is nothing that causes such prodigious and ponderous pollution as plastic. But the versatility of plastic materials is fantastic—it can be molded and laminated into any shape and for almost any application and lasts almost forever—everything from the clothes you wear, the car you drive, your phone, TV, computer, to buildings and space stations. Plastic is versatile, lightweight, flexible, moisture resistant, strong, relatively inexpensive, and has amazing staying power. The great irony is that its short-term use lasts many lifetimes.

In 1960, the world produced seven million tons of plastic. Consumption reached almost 300 million tons by the end of 2015. It's predicted to rise to 540 million tons by 2020. Of all the plastic produced, more than two-thirds has become plastic waste. Of that, only 9 percent has been recycled. The vast majority—79 percent—is accumulating in landfills, discarded as litter, much of it ending up in our waterways. In the next decade, more plastic will be produced than during the entire 20th century. It is estimated that about one trillion plastic bags are used worldwide every year—more than one million bags a minute, used and tossed within moments.

Think about the fact that just about every piece of plastic you have ever looked at, used, or thrown away still remains somewhere in the world today. The velocity and rate of plastic manufacturing has outpaced every other man-made material. This wonder product, which has transformed our lives in so many positive ways, has some serious environmental downsides in both its production and disposal, which threatens the health of the biosphere as well as our own.

At every stage of its life cycle it can be toxic because it's a petroleum product and unsustainable, producing greenhouse gases before, during, and after production and during disposal.

Everyone has heard of footprints of carbon, water, and energy, but the new and ubiquitous footprint is made of plastic. Because it is inexpensive, lightweight, and durable, virtually every industry loves it, but like a bad habit, it won't change or go away.

The Plastic Disclosure Project (PDP) is an attempt to change people's awareness and behavior toward plastic regarding its manufacture and usage, especially by businesses. They and other drum beaters are sounding the alarm over the amount of plastic that is used wastefully or that ends up as trash on land and in waterways. Many say that plastic pollution is now a major threat for the world's watercourses and for the global environment as a whole. By getting companies to assess and understand own their involvement in the problem, the project is hoping to raise awareness, and to change consumption and recycling habits.

Over 1,600 business organizations in the United States are involved in recycling postconsumer plastic items. One to 2,000 gallons of gasoline, around 2,000 pounds of oil—the amount of water used by one person in two months and the energy consumption for two persons for one year—can be saved by recycling just one ton of plastic waste.

Facts and figures:

- Each American throws away approximately 185 pounds of plastic annually.
- In the United States, we go through around 1,500 plastic water bottles every second.
- Seventeen million barrels of oil are used in producing bottled water each year. If we recycled three-quarters of the plastic pro-

duced annually, 1 billion gallons of oil and 44 million cubic yards of landfill space would be saved.

- In the United States, more than 60 million plastic bottles are thrown away every day and 80 percent wind up in a landfill, on land, or in the ocean.
- Five hundred million straws are used daily in the United States. Straws are among the top ten pieces of plastic found on beaches.
- Fifteen thousand plastic straws were picked up in the San Diego 2017 Beach Clean-Up.
- Seattle, Washington, and Manhattan Beach and Monterey, California, have banned straws.
- Miami Beach has banned plastic straws on the beach.
- In Davis, California, a new law requires a request for a straw.
- San Luis Obispo, California, enacted plastic straw restrictions.
- Wetherspoons, a restaurant chain, will replace plastic straws with paper—saving 70 million plastic straws per year.
- Museum Hotels is planning to reduce waste by cutting out 735,000 straws a year.

Plastic chemicals can be absorbed by the body—93 percent of Americans age six or older test positive for BPA (a plastic chemical). Many products in direct contact with foods and beverages use BPA (an estrogen-like compound), which could pose some risk to the brain development of babies and children. It may cause things like early puberty in girls and breast development in boys.

According to the industry website, a consumer would have to ingest more than 1,300 pounds of food and beverages in contact with polycarbonate plastic every day for an entire lifetime to exceed the safe level of BPA recently set by U.S. government agencies.

There is also some concern that there might be a possible link between BPA and increased blood pressure. Look for products labeled as BPA free. You can also reduce your exposure by limiting canned foods because most cans are lined with BPA. Use glass, porcelain, or stainless steel containers for hot foods and liquids instead of plastic containers. The National Institute of Environmental Health Sciences advises against microwaving polycarbonate plastics or putting them in the dishwasher because the plastic may break down over time and allow BPA to leach into foods.

PETRO-PLASTIC PROBLEMS

Twelve percent of the world's oil production goes to making plastic, of this, 4 percent is needed to power the plants that produce the plastic. Producing and recycling plastic requires transportation, generating more petroleum waste and emissions. Potentially hazardous chemicals are added during the production process and the risks posed by these additives are either barely tested or not tested at all. When plastic finally does break down, it releases a whole slew of toxic chemicals.

Styrene is a clear, colorless liquid that is derived from petroleum and natural gas by-products, is considered toxic, mutagenic (capable of changing genetic material), and possibly carcinogenic. Plastics contain phthalates (high content of which has been associated with asthma, breast cancer, type 2 diabetes, and fertility issues) and vinyl chloride (the most toxic plastic chemical linked to cancer, birth defects, and serious chronic diseases). Antimony is a chemical found in polyethylene terephthalate (PET) plastic bottles. In small doses, it can cause dizziness and depression; in larger doses, it can cause nausea, vomiting, and death. Finally, dioxin (a highly toxic compound produced as a by-product in manufacturing processes, can cause reproductive and developmental problems, damage the immune system, interfere with hormones, and also cause cancer). All of these chemicals end up in our air and water. It should be noted that it would take large doses, or prolonged exposure to them to cause serious illness.

Coding

To combat the recycling problem posed by the need to separate different plastics, many manufacturers have adopted a coding system. Containers are stamped generally on the bottom with a code indicating the type of plastic from which they are made. Coding makes it possible to sort containers in the recycling process. Codes are read manually from one to seven.

1. PET initially gained widespread use as a wrinkle-free fiber commonly called "polyester," and the majority of its production still goes toward textiles. It has become extremely popular for food and drink packaging because of its ability to create a liquid and gas barrier that keeps drinks fizzy and aids in keeping food fresh.

2. High-density polyethylene (HDPE) is a versatile polyethylene polymer that has the simplest basic chemical structure of any plastic, making it very easy to process and popular for packaging. However, it is rarely recycled because it is difficult to do on an industrial scale and it contaminates the recycling stream.

3. Polyvinyl chloride (PVC) is the second-most widely used plastic resin in the world. Use has decreased because of serious health and environmental pollution. It's rarely recycled because it is difficult to do on an industrial scale, and it contaminates the recycling stream.

4. Low-density Polyethylene (LDPE) is used in low-value products. Properties are strength, toughness, flexibility, resistance to moisture, ease of sealing, and ease of processing.

5. Polypropylene (PP) is slightly harder and more heat resistant than polyethylene and is used in a wide variety of applications, such as packaging.

6. Polystyrene (PS) is Styrofoam food containers and packing peanuts, foamed, and puffed with air. Weak in strength. Very low recycling rates.

7. Plastic #7. This category does not identify one particular plastic resin. It is a general catch-all and can include plastics that may be layered or a mixture of various plastics; it includes new bioplastics.

Micro-mess

Even though a lot of plastic waste is visible from roads to beaches, what is less visible is microplastics that get worked into clothing, cosmetics, cleaners, water, and us. Clothing made from synthetic material releases millions of fibers during washing that are so small they pass through filters in sewage treatment plants.

The American organization 5Gyres, a nonprofit that works against pollution, counted the microplastics in one single toiletry cleaner and found 360,000 pieces of plastic that flow from your drain straight out to sea and into marine plants and animals.

Look for labels that contain polyethylene, polypropylene, poly-e-terephthalate, and polymethyl methacrylate. Go to: http://www.businessinsider.com/products-that-contain-microbeads-2015-5, for a list of products containing microbeads.

Conventional perfumed room deodorizers not only come in plastic packaging, but they're also filled with hidden toxins. Essential oils work just as well without being noxious. Add a few drops of your favorite scent on the inside of the cardboard toilet paper roll, or put a few drops mixed with water in a glass jar and add some natural wooden dowels to make an aroma diffuser.

Soap nuts are a wonderful nontoxic, environmentally friendly alternative to conventional laundry and dish detergent. The nut contains saponin, a natural detergent that is released when they absorb water. One single soap nut can be used for 10 loads of laundry. You can even put three or four in a muslin bag and drop them in the utensil rack of your dishwasher for cleaning dishes. Not only do soap nuts replace plastic detergent jugs, but they can also be composted when you're done with them, and they are reasonably inexpensive

Bamboozled by Water Bottles

Bottled water started to take off around 1977 when the iconic green-glass Perrier bottle was introduced and became a status symbol. A rumor raced around that it was healthier than water out of the tap, a fable that some still believe. Big changes came in 1989 with a technological innovation—lightweight and inexpensive water bottles. Hundreds of millions of dollars were spent on marketing by big hitters Pepsi and Coca-Cola and many others telling us that their water was pure, healthy, clean, and crisp. It made ordinary tap water sound like hand-me-down clothes. Actually, U.S. water-quality standards set by the EPA for tap water are more stringent than for bottled water, and 64 percent of brand-name waters (like Coke and Pepsi, Aquafina, and Dasani) come out of municipal taps.

We use roughly 60 percent of the world's water bottles, even though we're less than 5 percent of the world population, and we throw away 50 billion plastic water bottles every year, the majority of which aren't recycled. Although approximately 800,000 tons of plastic bottles were recycled in 2011, more than twice as much was thrown away haphazardly.

On average, Americans spend about $100 per person each year on bottled water. Experts predict annual revenues to increase to $195 billion in 2018. At least 90 percent of the price of a bottle of water is for things other than the water itself, like bottling, packaging, shipping, and marketing. It takes up to 2,000 times as much energy to produce and transport the average plastic bottle of water to your home as it does to produce the same amount of tap water, and overall costs 2,000 times as much as tap water.

By the way, if you're really into conspicuous consumption, try a bottle of Acqua di Cristallo water that comes in a 24-karat solid-gold bottle and contains a small sprinkling of gold dust, at $60,000 per 750 milliliters. Chances are, unless you are a Kardashian or a Trump, you won't be tossing that bottle out.

The environmental consequences of the manufacturing, transport, and disposal of the bottles are brutal. The Earth Policy Institute estimates that making bottles to meet the U.S. demand requires enough oil to fuel 100,000 cars for a year, enough to lubricate a million cars and to power 190,000 homes annually. Transport and disposal of the bottles adds to the resources used, and water extraction adds to the strains bottled water puts on our ecosystem as simple waste.

Through campaign contributions, high-powered lobbyists, and expensive public relations campaigns, the plastics industry is able to keep proposed container deposit legislation bottled up in committees at both the state and national levels. They do not want the hassle that they went through with glass bottles.

- The recommended eight glasses of water a day, at U.S. tap rates is about $49 per year. The same amount of bottled water is closer to $2,000 reported by the *New York Times*.
- Reusing one bottle continually will prevent the purchase, consumption and waste of more than 240 plastic bottles every year, according to greenworldcitize.com.
- It takes about 1.39 liters of water to make 1 liter of bottled water, which means that a substantial amount of H_2O goes down the drain during production. That doesn't include the amount of water that's used to make the bottles themselves.
- San Francisco, California, joined Concord, Massachusetts, in banning the sale of bottled water within owned city property.

POSSIBILITIES

Bioplastics

Bioplastics are plastics made from biomass sources, meaning renewable sources such as vegetable oils. They seem to have great potential benefits but also an almost equal number of drawbacks. Regrettably, many biodegradable plastics may not biodegrade rapidly enough under needed environmental conditions to replace conventional plastic.

Pros:

- The biodegradable manufacturing process can use up to 65 percent less energy and generates fewer greenhouse gases than conventional plastic.
- They are supposedly biodegradeable and/or compostable, or can be recycled alongside conventional plastics.
- The market for bioplastics is growing by 20 to 30 percent a year. This will drive technological advances and lower the price of the product.

Cons:

- They have a higher manufacturing cost—though this is changing as more companies begin to make and use bioplastics.
- Recycling and composting may be possible only in industrial composting processes.
- Some can interfere with or damage standard plastic recycling processes.
- If sent to a landfill, some can release methane into the atmosphere, as they are plant sourced.
- They're not suitable for use in a number of products.
- Use of plant sugar and starch sources as raw material could have a negative impact on food prices.

There is also interest and concern over the growing use by supermarkets of so-called sustainable oxy-degradable plastic bags. They are made of conventional oil-based plastic, with an additive that enables them to break down far more easily. Manufacturers claim they release no methane or harmful residues, and rely on additives to the resin that hastens degradation upon exposure to different conditions.

The verdict about biodegradable plastics still seems to be out. *Earth News* claims that while some of these bioplastics are claiming to be "100 percent compostable," testing at Woods End Labs has found that most of these claims are misleading at best.

One user claims that the Woods End report stated that the bioplastic bags aren't designed to break down under most composting conditions. But Eco-Oxo, the manufacturer, says that the Oxo bags will break down when exposed to oxygen and heat.

The environmental impact of bioplastics is up for debate, as there are many different ways to determine "greenness" such as water use, energy use, deforestation, biodegradation, and other tradeoffs. These considerations are also complicated by the fact that many different types of bioplastics exist, each with different environmental strengths and weaknesses, so not all bioplastics can be treated equally.

The production and use of bioplastics is sometimes regarded as a more sustainable activity when compared with plastic production from petroleum resources because it requires less fossil fuel for its production and there are fewer greenhouse emissions (maybe net-zero) when it biodegrades, as it is made from plant sources that absorb CO_2 when growing. Bioplastics may make carbon savings of 30 to 80 percent compared with conventional oil-based plastics. However, concern is mounting because when the new generation of biodegradable plastics ends up in landfills and degrades without oxygen, it releases methane.

Refundable Recycling

Refundable deposits have always provided a financial incentive to return beverage containers for recycling. A report by Businesses and Environmentalists Allied for Recycling (BEAR) found that approximately 28 percent of the U.S. population that lived in states with a container deposit law recycled 490 containers per capita, as opposed to consumers in nondeposit states, who recycled only 191 containers per capita.

The U.S. recycling rate for PET bottles, which include food and nonfood bottles and jars, and all beverage bottles (except carbonated drinks), was only about 17 percent, while the PET soda bottle recycling rate was twice as much—34 percent.

What You Can Do

Educate yourself about the full cost of plastics and inform family and friends of ways that they can cut down on waste. Wean yourself off plastic bags. Over one trillion plastic bags are used worldwide every year, and the average American throws out 10 plastic bags a week. Carry reusable bags wherever you go. Break up with bottled water and bring your own refillable drink bottle, coffee cup, or thermos instead of buying disposable, single-use plastic cups and bottles.

Some coffee shops even offer customers a discount if they bring their own container. Ditch plastic lighters and use matches instead. Invest in an environmentally friendly razor that allows you to replace the head and not the whole razor, the plastic body of which is very tough to recycle—and two billion of them get tossed out every year.

Try to buy products that have less plastic packaging, or better yet, opt for ones that come in cardboard or other recyclable materials. Support companies who are trying to reduce their packaging.

Instead of wrapping your sandwich and snacks in plastic wrap, pack them in reusable containers, or nonbleached brown paper. When possible, buy products that can be refilled rather than those that are single use. Refuse straws. Buy products in bulk and bring your own reusable bags for shopping. Chewing gum actually contains plastic—and ABC gum (already been chewed) is impossible to recycle.

Replace all disposables with reusables. You probably don't even realize how many disposable items you use every day. Every piece of plastic you throw out has the potential to wind up in a river, lake, or ocean—forever.

If you're on the beach or in a park, leave only a footprint. And it doesn't hurt to pick up after your neighbors. Volunteer at a beach cleanup, they are a great way to help the environment and meet like-minded individuals who want to reduce their pollution footprint. Check out Surfrider Foundation chapters for monthly cleanups.

Remember and implement the five Rs—Refuse, Reduce, Reuse, Recycle, and Right on. A little effort goes a long way.

SOURCES

"6 Pros and Cons of Bioplastics." Health Research Funding, June 25, 2015, https://healthresearchfunding.org/6-pros-and-cons-of-bioplastics/.

Badore, Margaret. "11 Easy Ways to Reduce Your Plastic Waste Today." Treehugger, March 2, 2015, https://www.treehugger.com/green-home/11-easy-ways-reduce-your-plastic-waste-today.html.

Bauer, Brent A. "What Is BPA, and What Are the Concerns about BPA?" Mayo Clinic, https://www.mayoclinic.org/healthy-lifestyle/nutrition-and-healthy-eating/expert-answers/bpa/faq-20058331.

"Bioplastics May Not Be as 'Sustainable' as Claimed." *China Daily*, April 29, 2008, http://www.china.org.cn/environment/news/2008-04/29/content_15033439.htm.

"Boot the Water Bottle!" Pure Earth Technologies, http://www.pure-earth.com/bootthebottle.htm.

Brodwin, Erin, and Ashley Lutz. "Your Face Wash Could Contain an Ingredient That's Killing Fish and Turtles." *Business Insider*, May 26, 2015, http://www.businessinsider.com/products-that-contain-microbeads-2015-5.

Chow, Lorriane. "Plankton Eating Plastic Caught on Camera for First Time Ever." Eco Watch, July 9, 2015, https://www.ecowatch.com/plankton-eating-plastic-caught-on-camera-for-first-time-ever-1882064585.html.

D'Alessandro, Nicole. "22 Facts about Plastic Pollution (and 10 Things We Can Do about It)." EcoWatch, August 7, 2014, https://www.ecowatch.com/22-facts-about-plastic-pollution-and-10-things-we-can-do-about-it-1881885971.html.

"High-Density Polyethylene (HDPE)." Plastics Make It Possible, https://www.plasticsmakeitpossible.com/about-plastics/types-of-plastics/professor-plastics-high-density-polyethylene-hdpe-so-popular/.

"Down the Drain." *Waste Management World*, February 6, 2006, https://waste-management-world.com/a/down-the-drain.

Engler, Sarah. "10 Ways to Reduce Plastic Pollution." NRDC, January 5, 2016, https://www.nrdc.org/stories/10-ways-reduce-plastic-pollution.

"ESG Wrap." *Boardroom.Media*, March 16, 2018), https://boardroom.media/articles/single/esg_wrap_|_gen_y_brand_dump_|_blockchain__tuna_|_plastic_waste_|_uk_gig_economy_10102.

Faraz, Sabrina. "Green Media Initiatives." Strengthening Environmental Journalism, March 4, 2018, https://gminitiatives.wordpress.com/2018/03/04/end-plastic-pollution-peacewalk/

"FDA Bottled Water Regulations." International Bottled Water Association, http://www.bottledwater.org/education/regulations/fda-vs-epa.

Franklin, Pat. "Plastic Water Bottles Should No Longer Be a Wasted Resource." Container Recycling Institute, http://www.container-recycling.org/index.php/issues/.../275-down-the-drain.

Good, Kate. "10 Life Hacks to Help You Cut Plastic Out of the Picture." *One Green Planet*, December 31, 2014, http://www.onegreenplanet.org/environment/life-hacks-to-help-you-cut-plastic-out-of-the-picture/.

Gourmelon, Gaelle. "Global Plastic Production Rises, Recycling Lags." *Vital Signs*, January 27, 201, http://www.worldwatch.org/global-plastic-production-rises-recycling-lags-0.

"Guinness Record Goes to Highest Price Water Bottle: Acqua di Cristallo Tributo a Modigliani." *Inventor Spot*, http://inventorspot.com/articles/guinness_record_goes_highest_price_water_bottle_acqua_di_cristal_39184.

"Highlights of Low-Density Polyethylene." Plastics Make It Possible, https://www.plasticsmakeitpossible.com/about-plastics/faqs/professor-plastics/professor-plastics-highlights-of-low-density-polyethylene/.

"How Can I Tell What Type of Plastic Something Is Made of, and If That Plastic Is Safe?" Life Without Plastic, https://www.lifewithoutplastic.com/store/common_plastics_no_1_to_no_7#.WuDzuTNlDcs.

"How Did Bottled Water Get So Popular?" *Filter Butler*, http://filterbutler.com/blog/how-did-bottled-water-get-so-popular/.

Intagliata, Christopher. "Does Recycling Plastic Cost More Than Making It?" *Live Science*, November 3, 2012, https://www.livescience.com/32231-does-recycling-plastic-cost-more-than-making-it.html.

"It's Cheap . . . It's Disposable . . . It's Toxic." The World Counts, July 1, 2014, http://www.theworldcounts.com/stories/What-is-Plastic-Pollution.

Lake, Rebecca. "Bottled Water Statistics: 23 Outrageous Facts." *Credit Donkey*, July 11, 2015, https://www.creditdonkey.com/bottled-water-statistics.html.

Leblanc, Matt. "Plastic Recycling Facts and Figures." The Balance Small Business, June 10, 2017, https://www.thebalancesmb.com/plastic-recycling-facts-and-figures-2877886.

Messinger, Leah. "How Your Clothes Are Poisoning Our Oceans and Food Supply." *Guardian*, June 20, 2016, https://www.theguardian.com/environment/2016/jun/20/microfibers-plastic-pollution-oceans-patagonia-synthetic-clothes-microbeads.

Moss, Laura. "16 Simple Ways to Reduce Plastic Waste." Mother Nature Network, https://www.mnn.com/lifestyle/responsible-living/stories/16-simple-ways-reduce-plastic-waste.

Naik, Sitaram. "Simple solutions for Complex Problems: Plastic Waste." Complex Problems, http://www.complexproblems.in/Plastic-waste.htm.

"Now Accepting #1–7 Plastic Bottles, Tubs and Jars in Your Curbside Recycling Bin!" Eco-Cycle, http://www.ecocycle.org/plastics-recycling.

Nunez, Jackie. "The Sipping Point." The Last Plastic Straw, https://thelastplasticstraw.org/about-us/.

Parker, Laura. "Plastic Ain't So Fantastic." *Ocean Crusaders*, July 19, 2017, http://oceancrusaders.org/plastic-crusades/plastic-statistics/.

Pereda, Otto. "4 Tips to Reduce Plastic Pollution." *Costa Rica News*, March 27, 2018, https://thecostaricanews.com/47912-2/.

"Plastic." Alburgh Transfer Station, http://alburghtransferstation.com/plastic.html.

"Plastic Water Bottle Pollution." Live Life Healthy, http://waterbottles.healthyhumanlife.com/plastic-water-bottle-pollution-plastic-bottles-end/.

"Polystyrene." Chemical Safety Facts, https://www.chemicalsafetyfacts.org/polystyrene-post/.

"Polyethylene terephthalate." *Encyclopedia Britannica*, https://www.britannica.com/science/polyethylene-terephthalate.

"Prince's Award for Innovative Philanthropy." Plastic Disclosure Project, January 25, 2018, http://plasticdisclosure.org/.

Lamb, Robert. "Bioplastics: The Pros and Cons." How Stuff Works, https://science.howstuffworks.com/environmental/green-science/corn-plastic2.htm.

Schriever, Norm. "Plastic Water Bottles Causing Flood of Harm to Our Environment." *Huffington Post*, December 6, 2017, https://www.huffingtonpost.com/norm-schriever/post_5218_b_3613577.html.

"Soap Nuts for Natural Laundry Care." Wellness Mama, April 3, 2018, https://wellnessmama.com/7553/soap-nuts-laundry-care/.

"Plastic Pollution—Facts and Figures." Surfers against Sewage. https://www.sas.org.uk/our-work/plastic-pollution/plastic-pollution-facts-figures/.

"The Facts about Bisphenol A." WebMD, https://www.webmd.com/children/bpa#1.

"Bisphenol A (BPA): Use in Food Contact Application." U.S. Food and Drug Administration, November 2014, https://www.fda.gov/NewsEvents/PublicHealthFocus/ucm064437.htm.

"The Last Plastic Straw." Plastic Pollution Coalition, http://www.plasticpollutioncoalition.org/no-straw-please.

"The Past, Present and Future of Single-Use Plastic Bags." *Ivie*, September 18, 2014, https://www.science.org.au/curious/earth-environment/future-plastics.

"These Easy Things Can Dramatically Decrease the Amount of Plastic Trash You Make." Complex Problems, July 29, 2015, http://www.complexproblems.in/Plastic-waste.htm.

Vaughn, Aubrey. "Oxo Biodegradable Bag Test." *Mother Earth News*, August 18, 2010, https://www.motherearthnews.com/nature-and-environment/oxo-biodegradable-bags-test-1.

Vidal, John. "Sustainable' Bio-Plastic Can Damage the Environment." *Guardian Weekly*, April 25, 2008, https://www.theguardian.com/environment/2008/apr/26/waste.pollution.

Wassinger, Bettina. "Raising Awareness of Plastic Waste." *New York Times*, August 14, 2011, https://www.nytimes.com/2011/08/15/business/energy-environment/raising-awareness-of-plastic-waste.html.
"What Are Microplastics?" National Ocean Service, https://oceanservice.noaa.gov/facts/microplastics.html.
"What Is BPA? Should I Be Worried about It?" Drugs.com, https://www.drugs.com/mcf/what-is-bpa-should-i-be-worried-about-it.
"What to Do With Plastic Waste?" Plastic Soup Foundation, https://www.plasticsoupfoundation.org/en/files/what-to-do-with-plastic-waste/.

11

PACKAGING

Thinking Inside and Outside the Box

Packaging is the planned excess and obsolescence of things

Who hasn't experienced "wrapping rage," growling, cursing, slashing, pulling, tearing, and its sidekicks, anger and frustration, resulting from trying to dislodge some desired doodad from the bondage of cardboard and heat-sealed plastic blister packs, clamshells, or other confined encasements. What's worse is that overpackaging mania represents a third of waste thrown away in America. And less than 14 percent of plastic packaging, the fastest-growing form of packaging, gets recycled. Packaging adds 29 million tons of nonbiodegradable waste to landfills every year.

According to LiquiGlide Inc. (a company that creates slippery, liquid-impregnated surfaces developed at MIT) a survey of more than 1,000 consumers, people would rather go to the dentist than waste consumer products they have purchased. And while that's largely about the money, one-sixth also cited environmental concerns.

A cautionary tale: Think about a waterfront restaurant owner on a Thai island whose customers used to wrap their lunch in banana leaves, which they would then casually toss in the surf when done. That was OK because the leaves peacefully decomposed and the fish ate the scraps. But in the past decade, plastic wrap rapidly replaced banana leaves, leaving the beach in a crust of plastic.

BAGGING THE PLASTIC BAG

Americans use 100 billion plastic bags a year, which require 12 million barrels of oil to manufacture. Of all the plastic problems people face, the aggravating single-use plastic bag is the baddie. There was recently a kind of test case for bagging the problem in California. That state has been in the lead for many ecological causes in the past. Proposition 67, a plastic bag law, was passed in 2016.

Currently, single-use plastic bags at large retail stores have been banned in the state. Until this year, every Californian, on average, used about 400 plastic bags a year, forcing the state to spend $500 million in annual litter cleanup, with between 8 percent to 25 percent attributable to plastic bags.

And the ban seems to be working. In San Jose, the storm drain systems are 89 percent cleaner and streets and creeks have been reported as 60 percent cleaner. In Los Angeles County, the ban resulted in a 94-percent reduction in single-use bag use. In Alameda County, officials reported finding 433 plastic bags, compared to 4,357 in 2010. Monterey County reported even better news, with volunteers discovering only 43 plastic bags while performing their clean-up efforts, compared to 2,494 in 2010.

So either people are so cheap that they are using fewer bags because they have to buy them, or they're getting the message about plastic pollution, or a little of both.

Facts and figures:

- Packaging makes up 30 percent of the weight and 50 percent of municipal solid waste (MSW) trash by volume.
- The Clean Water Act found that 49 percent of litter is fast-food packaging products.
- Containers and packaging alone contribute more than 23 percent of the material reaching landfills in the United States.
- Packaging adds more than 30 million tons of nonbiodegradable waste to landfills every year.
- According to the EPA, over 380 billion plastic bags, sacks and wraps are consumed in the United States each year.
- Four out of five grocery bags in the United States are now plastic.

- The average family accumulates 60 plastic bags in only four trips to the grocery store.

Manufacturers, despite the threat to our ecosystem, still choose plastic over other material for containers for many reasons. Plastic containers are a lot cheaper. Plastic packaging takes up less space than glass, it's lighter, which is a benefit to manufacturers who are prone to buy in bulk, to retailers who have to transport them, or consumers that buy on the internet and have the product shipped to them. And it takes less energy to produce plastic than paper or glass.

REUSE TO USE

California has a goal of reaching a 75-percent recycling rate by 2020 that could create 110,000 jobs. So it's important for restaurants, beverage companies, and grocery stores to take responsibility for their packaging.

Then there's the question of recyclability. Both glass and plastic containers can be recycled, yet in reality glass is recycled less than plastic packaging. The Glass Packaging Institute notes that recycling glass uses 66 percent of the energy it would take to manufacture new glass on average, while plastic only requires 10 percent of the energy it takes to produce new plastic.

Punched-Out Plastic

Plastic juice pouches and drink boxes are generally not recyclable. The 1.4 billion Capri Sun pouches thrown away every year laid end-to-end would reach nearly halfway to the moon. The National Resources Defense Council (NRDC) joined the Make It, Take It campaign (a coalition of organizations devoted to waste recycling and resource conservation) to ask companies like Kraft, which produces and distributes Capri Sun, to use recyclable, reusable, or compostable packages for beverages. They have made comments about an effort to recycle, but when asked if they plan to improve their packages or efforts to improve the sustainability of their product, Capri Sun declined to comment.

Today, there are fundamentally four choices for companies that package consumer goods: nonrecyclable, recyclable, biodegradable, and glass, which seems to be disappearing from the market shelves. Biodegradable packaging is the latest step in sustainable packaging, and many believe it is the ultimate solution to the packaging waste problem. However, according to estimates there isn't enough land available to grow sufficient and suitable crops to substitute for traditional plastics. (See chapter 10, Plastics: A Blessing and a Curse.)

GETTING BOXED ONLINE

Eight in 10 Americans are now shopping online, accounting for about 10 percent of all retail sales, and just about all products arrive in cardboard boxes.

Every day, at the Recology plant in San Francisco, approximately 625 tons of recyclables, including more than 100 tons of cardboard, are collected, and the amount of plastic film has increased for eleven consecutive years. Plastic film recycling—a category that includes flexible product wraps, bags, and commercial stretch film made primarily from polyethylene (PE)—has increased nearly 84 percent since the first report was issued in 2005.

With the growing popularity of on-demand delivery and meal kits, more nonrecyclable packaging is making its way to the recycling center. Box liners, ice packs, some juice boxes, and other parts are not recyclable—despite the manufacturers' claims that they are.

A Diet of Detritus

Still, lots of small electronics like cables and storage cards are sold in heavy plastic packages, and for that very small article the packaging can be up to fifteen times larger than the actual item, while companion items are wrapped in separate pieces of plastic. Most packaging of small items (like makeup) comes inside a plastic container, which is then wrapped in more cardboard and plastic, adding up to more total packaging than the volume of the makeup itself.

The inflated size of a bag of chips or other snacks is not equivalent to the amount of food actually in the bags, which are puffed with air to look fuller and to protect the product, but the amount of packaging could be cut in half.

Over the last 20 years, e-commerce has evolved from something novel into a $300 billion plus business delivering products arriving in various states of packaging. Despite all this, the amount of cardboard shipped by U.S. companies has been decreasing slightly for the last 10 years, according to the Fiber Box Association.

The EPA also stated that by 2010, packaging waste should have grown by nearly a third to account for 38 percent of waste. Instead, it was actually reduced by 24 million tons annually. It looks as if packaging has gone on a continual diet from heavier, traditional materials to those that are stronger and lighter, such as plastics, multilayer films, and multimaterial cartons, because they are cheaper. Plastic bottles are significantly thinner and lighter than a few years back, and are easier to crush and recycle. Also packaging engineers are figuring out how to use as little material as possible.

One of the reasons for the decrease is that manufacturers have been improving their packaging to reduce the cost of shipping (and less greenhouse gases), a major factor in what shoppers buy. Some manufacturer may substitute corrugated trays shrink-wrapped in plastic, eliminating cardboard altogether.

Overall, thanks to source reduction and recycling, on a per capita basis, Americans reduced the percentage of packaging headed to landfills by an outstanding 39 percent—in fewer than 20 years.

DON'T CELEBRATE YET

Too much of what goes into recycling is "contaminated" and ends up in a landfill. The glue on bows, the glitter dusting on your fancy wrapping paper, dirty pizza boxes, and plastic grocery bags—clog the process of turning waste paper into new paper and cardboard, making recycling more of a hassle and extra costly.

China, our best customer for garbage, is setting new limits on the contamination it will allow in mixed paper bales that American trash companies ship for recycling. "They've started getting more rigorous, even tearing open bales at customs," said Chaz Miller, policy director for the National Waste & Recycling Association. Because the income from exporting our garbage to China often subsidizes the cost of some neighborhood garbage collection programs, taking less of America's used paper will cause our bills to collect, recycle, or dispose of it, to rise.

And guess what—many of the items we order online are made in China and come in recycled cardboard boxes from paper material bought from the United States.

WHAT YOU CAN AND CAN'T DO

Seasonal Setbacks

Yes, you can recycle Christmas wrapping paper—unless it's metallic or has glitter or velvety flocking on it. Plain wrapping paper is totally recyclable. And don't worry about getting all the tape off before you toss it in the bin. Opt for reusable gift bags that can be reused. Bows are the easiest gift wrap item to reuse rather than recycle. Even if they lose their stickiness, a bit of tape makes them as good as new.

Facilities that accept commingled recyclables sort out the cardboard using a large piece of equipment called disc screens, or sets of spinning discs with spaces in between. Large flat items like cardboard are carried up and over the screen, while smaller items fall through the spaces between the discs. Unfortunately, ribbons (as well as plastic bags, twine and anything else that's long and stringy) end up wrapping around the spinning shafts that hold the discs and muck with the works. This causes the facility to shut down so they can get in and cut out all the junk that has wrapped around the shafts holding the discs.

Plain paper cards can go straight into the paper recycling bin, no problem. But the shiny cards printed on photo paper need to go into the trash and it's the same for ones that have lots of metallic embossing, though usually you can tear the card in half and at least recycle a part. Cards that have a lot of glitter on them should also stay out of recycling.

Boxes

Cardboard is a great material to recycle because, generally, it's clean and easy to reprocess—and every ton of it that's reclaimed saves 17 trees. The most important thing when recycling boxes is simply to break them down flat. Otherwise, they take up too much room in the recycling bin and on trucks requiring more energy for more trips creating more pollution.

Packages and boxes closed with paper tape don't require anything more than flattening. For heavy-duty, wide plastic tape, peel it off. While breaking down the boxes, be cheered that everything you're doing you'll have to do all over again when the cardboard comes back to you—possibly from China—with your new stuff.

Here are some other tips:

- Always carrying your own reusable bags while shopping.
- Get fruit and vegetables packed only in paper bags, rather than in plastic or on polystyrene trays.
- Buy food and drink in recyclable packaging made of glass jars or tin cans when you can.
- Purchase dry goods in bulk—this means fewer individual packages.
- Get basic ingredients and cook them yourself, rather than small prepackaged portions.

SOURCES

"Agri-Plast." Tapegear, http://www.tapegear.com/agriplastbw/.

"All That Online Shopping Has Cardboard Consequences." Waste Advantage, June 20, 2017, https://wasteadvantagemag.com/all-that-online-shopping-has-cardboard-consequences.

Belsey Priebe, Maryrut, "Reduce Packaging Waste." EcoLife, http://www.ecolife.com/recycling/tips-basics/reduce-packaging-waste.html.

"California Proposition 67, Plastic Bag Ban Veto Referendum." Balletpedia, 2016 https://ballotpedia.org/California_Proposition_67,_Plastic_Bag_Ban_Veto_Referendum_.

Cheeseman, Gina-Marie. "Fast Food Garbage Makes up 50% of Street (and Pacific Gyre) Litter." *Triple Pundit*, June 27, 2011, https://www.triplepundit.com/2011/06/fast-food-big-source-trash-pollution/.

Cho, Renee. "What Happens to All That Plastic?" *State of the Planet* (blog), January 31, 2012, http://blogs.ei.columbia.edu/2012/01/31/what-happens-to-all-that-plastic/.

Corkum, Kim. "10 Worst Examples of Packaging Waste." Plan, June 16, 2014, http://www.postlandfill.org/examples-of-packaging-waste/.

"Does Plastic or Glass Require More Energy to Recycle?" Wise Geek, http://www.wisegeek.com/does-plastic-or-glass-require-more-energy-to-recycle.htm.

"Coastal Clean-Up: California's Plastic Bag Ban Is Working." *Monterey Herald News*, November 14, 2017, http://www.montereyherald.com/article/NF/20171114/LOCAL1/171119928.

"Eliminate or Minimize Unnecessary Packaging." Rethink Recycling, https://www.rethinkrecycling.com/how-reduce-shipping-materials/eliminate-or-minimize-unnecessary-packaging.

"Environmental Stewardship Made Easy." *St. Augustine Record*, April 25, 2018, http://www.staugustine.com/opinion/20180425/record-editorial-environmental-stewardship-made-easy.

Kessler, Sarah. "The 20-Year Explosion in E-Commerce Has Not Increased US Cardboard Production." *Quartz Media*, February 17, 2017, https://qz.com/910620/the-20-year-explosion-in-e-commerce-has-not-increased-us-cardboard-production/.

Koenig, Seth. "Divided Portland Council Bans Polystyrene Containers, Adds Fees on Paper and Plastic Bags." *Bangor Daily News Portland*, June 17, 2014, http://bangordailynews.com/2014/06/17/news/portland/divided-portland-council-bans-polystyrene-containers-adds-fees-on-paper-and-plastic-bags/.

"Learn How Plastic Bags Are Made and Recycled. " Crown Poly, http://www.crownpoly.com/learning-center/.

LeBlanc, Rick. "Plastic Recycling Facts and Figures." *The Balance Small Business*, June 1, 2017, https://www.thebalancesmb.com/plastic-recycling-facts-and-figures-2877886.

Lehner, Peter. "Fast Food Trash Nation? Time to Cut Down on Packaging Waste." NRDC, February 6, 2015, https://www.nrdc.org/experts/peter-lehner/fast-food-trash-nation-time-cut-down-packaging-waste.

Lilienfeld, Bob. "From Crisis to Myth: The Packaging Waste Problem." *Live Science*, April 22, 2015, https://www.livescience.com/50581-packaging-no-longer-the-nightmare-some-claim.html.

Lingle, Rick. "Consumers Loathe Product Waste Because of Money, Mindset and the Environment." *Packaging Digest*, November 19, 2014, http://www.packagingdigest.com/optimization/consumers-loathe-product-waste-because-ofmoney-mindset-environment-141119.

Lucas, J. "Facts about Plastic Bag Use." *Post Consumers*, April 10, 2013, https://www.postconsumers.com/2013/04/10/plastic-bag-environmental-facts/.

Perez, Sarah. "79 Percent of Americans Now Shop Online, but It's Cost More Than Convenience That Sways Them." *TechCrunch*, December 19, 2016,https://techcrunch.com/2016/12/19/79-percent-of-americans-now-shop-online-but-its-cost-more-than-convenience-that-sways-them/.

"Plastic Bag Bans." Change.org, https://www.change.org/t/plastic-bag-bans-en-us.

"Plastic Bag Consumption Facts." Conserving Now, https://conservingnow.com/plastic-bag-consumption-facts/.

"Recycling Juice Boxes." Green Lifestyle Changes, December 6, 2009, http://www.greenlifestylechanges.com/recycling-juice-boxes.

Silverman, Ken. "Recycling Rates Are Rising for Plastic Bags and Wrap." *Environmental Leader*, March 7, 2017, https://www.environmentalleader.com/2017/03/recycling-rates-rising-plastic-bags-wrap/.

Sleight, Kenneth. "How Quickly Does Plastic Breakdown? The New Biodegradable Plastic Option." *Bright Hub*, February 21, 2011, https://www.brighthub.com/environment/green-living/articles/107380.aspx.

"Success! California's First-in-the-Nation Plastic Bag Ban Works." *Mercury News*, November 13, 2017, https://www.mercurynews.com/2017/11/13/editorial-success-californias-first-in-the-nation-plastic-bag-ban-works/.

Szaky, Tom. "Biodegradable vs. Recyclable: Which Is the Better Packaging Solution?" *TreeHugger*, December 4, 2008, https://www.treehugger.com/corporate-responsibility/biodegradable-vs-recyclable-which-is-the-better-packaging-solution.html.

"Taxpayer Cost of Plastic Bags in California." Green Eco Services, January 29, 2018, http://www.greenecoservices.com/taxpayer-cost-of-plastic-bags-in-california/.

Tellus Industries, *From Waste to Jobs: What Achieving 75 Percent Recycling Means for California*. NRDC Report. Boston: Tellus, 2014, https://www.nrdc.org/sites/default/files/green-jobs-ca-recycling-report.pdf.

"The Advantages and Disadvantages of Plastic Packaging." Packaging Innovation, June 30, 2014, https://www.packaginginnovation.com/packaging-materials/plastic-packaging-2/advantages-and-disadvantages-plastic-packaging/.

"The Misleading Biodegradability of PLA." *Filabot* (blog), October 23, 2015, https://www.filabot.com/blogs/news/57233604-the-misleading-biodegradability-of-pla.

"U.S. Cardboard Production Has Decreased Despite Decades of E-Commerce Growth." *Quartz Media*, February 17, 2017, https://www.sdcexec.com/sourcing-procurement/news/12306927/us-cardboard-production-has-decreased-despite-two-decades-of-ecommerce-growth.

Weise, Elizabeth. "China Is Setting New Limits on the Contamination It Will Allow in Mixed Paper Bales American Trash Companies Ship for Recycling." *USA Today*, December 25, 2017, https://www.usatoday.com/story/tech/news/2017/12/20/dont-recycle-bow-sloppy-christmas-recycling-can-send-all-your-efforts-landfill/963415001/.

Weise, Elizabeth. "You've Been Recycling Gift Wrap Wrong. Follow These Tips to Get It Right." *US Today*, December 20, 2017, https://www.king5.com/article/news/nation-world/youve-been-recycling-gift-wrap-wrong-follow-these-tips-to-get-it-right/507-500966533.

12

E-WASTE

The Stuff We Couldn't Do Without Is Now Stuff We Don't Know What to Do With

> Yep, it's revolutionary. We made a new iPad at twice the price and half the size, and durability. Also, if you break it, it's okay. We'll have six more models by next month.
>
> —*Jack Wynn*

Our electronic devices have become more than just machines—they have become digital extensions of us, and to lose them is almost akin to losing a limb. One problem is that we have an addiction to bleeding-edge technology to make our lives easier, quicker, and more efficient, but with that comes a price. Piles and piles of old devices that have limited life spans have become masses of electron waste—the toxic legacy of the automated age. We change out our gadgets almost as often as we buy new clothes.

Electronic waste, or e-waste, describes discarded electronic devices, which too often are not designed for efficient recycling or disposal. Other problems driving the e-waste crisis include few financial incentives to recycle, compulsory laws that regulate the disposal of e-waste, and the manufacturers reluctance to factor in responsibility for end of life planning.

While we were busy planning, inventing, installing, and becoming the world leaders in technology, we forgot an important part of our brave new world of techno evolution—what to do with the leftovers. The EPA reports that in 2013 the average U.S. household contained 28 consumer electronics, many of which were no longer being used because of obsolescence.

The United States and China generated the most e-waste, about 32 percent of the world's total. However, America recently lost the title of e-waste heavyweight to China. Each year together they generated the weight of 200 Empire State Buildings or 11 Great Pyramids of Giza. China is also the largest importer of e-waste and is home to the world's largest e-waste dumpsites.

However, on a per capita basis, some other countries famed for their environmental awareness and recycling records lead the way. Norway is on top of the world's electronic waste mountain, generating 62.4 pounds per inhabitant. At present, America generates about 11.7 million tons per year, or upward of 26 pounds per capita, according to a National Public Radio report.

Of all the e-waste generated in the United States in 2013, approximately 60 percent went into landfills and incinerators and only 1.27 million tons was recovered for recycling. A significant amount of the leftover of approximately 40 percent was exported.

Built-in obsolescence continues to have us replacing old, sometimes not even defective, appliances, rising from 3.5 percent in 2004 to 8.3 percent in 2012. The share of large household appliances that had to be replaced within the first five almost doubled from 2004 to 2013. A large part of the problem is the manufacturers. They have given us fewer ways to keep older electronics functioning effectively—now it's just toss out the old and buy something new. For this reason, reducing e-waste cannot fall on the shoulders of consumers alone. One possible alternative is to require producers of electronics to offer buy-back or return programs for old equipment. Or companies that effectively process old equipment could be given tax breaks.

E-WASTE EXPOSED

The life span of a cell phone is now only 18 months and a laptop is around two years, which shouldn't be a shocking revelation. But what is alarming is how little the public knows about the immensity of e-waste or the proper ways to dispose of it. Of the total electronic waste generated, only about 25 to 29 percent is recycled.

Our dirty little secret is that when your electronic waste goes to a recycler, about 80 percent of that material very quickly finds itself on a container ship delivering waste to a country like China, Nigeria, India, Vietnam, or Pakistan—where the process of recycling is kind of nightmarish. Workers on e-waste dumps are paid an average of $1.50 per day and are unprotected while working with toxic substances that are buried or burned in or near landfills close to their homes, according to the Electronic Revolution E-Waste website.

Facts and figures:

- Humans generate around 52 million tons of electronic waste every year, worldwide. That's like throwing away 800 laptops every second.
- E-waste comprises 70 percent of our overall toxic waste.
- The United States gets rid of 142,000 computers and more than 416,000 mobile devices every single day, according the EPA.
- Recycling one million laptops saves the energy equivalent of the electricity used by 3,657 U.S. homes in a year, according to the EPA.
- For every one million cell phones that are recycled, the EPA states that 35,274 pounds of copper, 772 pounds of silver, and 75 pounds of gold, worth $60 million are salvaged. In addition, there are 33 pounds of palladium, a precious metal used for making electrical contacts.
- One ton of circuit boards are estimated to contain up to 800 times more gold than one metric ton of ore. There is 30 to 40 times more copper in a ton of circuit boards than can be mined from one metric ton of ore.
- Compared to disposal in landfills or by incinerators, reusing or recycling computers can create 296 more jobs per year for every 10,000 tons of computer waste processed.
- Old television sets as well as CRT (cathode ray tube) monitors contain approximately four to eight pounds of lead, a neurotoxin. Improper disposal means this toxic substance can leach into the ground, and can find its way into the water table.

- Three hundred million computers and one billion cell phones go into production annually. These numbers are expected to grow by 8 percent per year.
- Seventy percent of our e-waste is sent to landfills and incinerators. Most of it is burned, releasing harmful toxins in the air.
- The most common hazardous electronic items include LCD desktop monitors, LCD televisions, plasma televisions, and older TVs and computers with CRTs.

REDUX OR RECYCLE

Many major retailers will take e-waste for recycling, regardless of whether you purchased the product from them or not. Always call ahead of time to confirm that stores will accept e-waste, what types of products they will recycle, and if there will be any fees.

Not all e-waste recyclers are the same. There are safe ways to recycle e-waste, and then there are companies that simply export the waste to developing countries with lax recycling laws. Look for an e-waste recycling company that has been vetted through e-Stewards (http://e-stewards.org/), a safe and reliable electronics waste recycling standard created by the Basel Action Network (BAN), a charitable nongovernmental organization working to combat the export of toxic waste. According to BAN, the United States is the only developed country that has not banned selling its e-waste to other countries.

Our electronic waste poses a real threat to our ecosystem by contaminating our water, air, and soil with toxic materials including lead, mercury, arsenic, cadmium, and other substances harmful to humans and wildlife, plus the waste is filled with all sorts of plastics and metals, making the separation process difficult.

So, with a short useful life, devices fall into e-waste at a rapid pace. In fact, it was estimated that there were 422 million unused and unwanted cell phones just accumulating in people's homes by the end of 2015.

Dump the Devices?

Just because devices are older doesn't have to mean the gadgets are toast. They can be donated, sold, refurbished—that's not done enough. Educate yourself on alterative possibilities other than the dump. Search recycle databases for outlets in your area and:

- Pass them on for reuse.
- Recycle.
- Learn to fix broken gadgets.
- Trade-in for cash, or trade on Gazelle.com.
- Sell them on the internet on sites like Craigslist, Amazon, eBay, sliBuy.com, or the many other sites.
- If it's broken or just too old to sell or donate, dispose of it safely.
- For TVs, call a waste disposal company to find a recycling site. Most states have laws preventing you from leaving them and other electronics outside for trash pickup.

Printer Cartridges

According to Cartridge Fundraising, more than 13 print cartridges are thrown away every second in the United States The proper way to dispose of them is to take them to your retailer, ship them back to the manufacturer, find a refill location, refill them yourself, drop them off free of charge to one of over 1,600 FedEx Offices, or dispose of them safely at your municipal hazardous waste site.

CDs

There is a limited recycling market for plastic in these forms because of the types and combinations of plastic (numbers 1, 5, 6, and 7). But instead of throwing them away, consider the following:

- It is estimated that it will take over one million years for a CD to completely decompose in a landfill.
- Every month approximately 100,000 pounds of CDs become obsolete.

- Donate old CDs, DVDs, and tapes to a secondhand store or music reseller for reuse. Even if the items are scratched, it's likely they can be repaired and resold.
- Use them for a DIY art project, or in the garden to scare birds off your fruits and veggies.
- Mail to a company like the CD Recycling Center of America or GreenDisk. Be sure to remove any paper sleeves.
- Find a drop-off location for CDs and tapes near you using the Recycling Locator.

Phone It In

There are more cell phones than people in the United States. Over 135 million cell phones were trashed in 2010 alone. According to the Channel Pro Network, 70 million cell phones are lost each year and only 7 percent are recovered. Cell phones have the highest recycling market of any electronic material. If the cell phone battery is removable, you'll want to take it out. Most cell phones have lithium-ion batteries, which require special transportation for recycling, so this is especially important if using a manufacturer's mail-back program.

Unfortunately, most cell phones (like iPhones) are deliberately designed to make disassembly difficult. Yes, they do it on purpose. They don't want you to get your phone repaired; they want you to buy a new one. Check out Quora to share opinions and knowledge.

Manufacturers should sell products complete with prearranged recycling service or subscriptions that can exchange old units for new ones rather than throwing them away. Consumers need to demand better end-of-life options for our high-tech trash; and consumers should press for safe recycling, or repairing.

Monitor Your monitor

Many of these instruments contain lead. If you have a LCD screen, it's likely being backlit by small fluorescent bulbs containing mercury and metals that you want to keep inside the monitor and out of a landfill.

When it's time to replace your monitor, be careful:

- Unplug any cables or power cords from the monitor to lighten the load and to prevent tripping.
- Use a dolly or a rolling chair to transport the monitor to your car. CRT monitors can be quite heavy and you don't want to risk dropping them—and spilling lead.
- Place a towel or blanket down in the car and put the monitor face down. This will ensure that nothing breaks if you hit any potholes.

More than 70 percent of your smaller electronic devices can be recycled, recovering items such as plastic, steel, aluminum, copper, gold, and silver that can be used in new products.

Batteries

Americans go through three billion dry cell and 350 million rechargeable batteries every year. It is illegal in most places to throw batteries in the garbage, as the EPA considers them hazardous waste. Call2Recycle's recycling map offers a thorough rundown of the laws that apply to each state (https://www.call2recycle.org/recycling-laws-by-state/). For more options, try the recycling locator (https://www.recyclenow.com/about-recycling-locator), or earth911 (https://search.earth911.com).

Unlike single-use batteries, rechargeable batteries be charged about 1,000 times before they need to be replaced. When spent, they should be properly recycled.

TOXIC TROUBLES

The growing toxic nightmare that is e-waste is not confined to third world outposts. Some nations are refusing e-waste exports, exacerbating problems especially in the United States. And shockingly, it's not just third world countries that use cheap labor to dismantle e-waste. For years, U.S. federal prisons used inmates to recycle computers and other electronics in ways that violated health, safety, and environmental laws, according to a report by the Justice Department's Office of the Inspector General.

There are concrete steps the government, manufacturers, and consumers can take to better dispose of electronic trash to help prevent the pileup of more e-waste and mitigate the hazards e-waste processing poses.

Mercury-containing devices can be sent to safe recycling facilities. Circuit boards should be sent to specialized and accredited companies where they are smelted to recover nonrenewable resources such as silver, tin, gold, palladium, copper, and other valuable metals.

Hard drives are shredded and processed into aluminum ingots for use in automotive industry. Ink and toner cartridges are taken back to their respective industries for remanufacture while those that can't be used are separated into metal and plastic for reuse as raw materials.

Batteries are taken to specialized recyclers where they are hulled to take out plastic. The metals are smelted in specialized conditions to recover nickel, steel, cadmium, and cobalt that are reused for new battery production and fabrication of stainless steel.

Currently, 25 U.S. states, covering 65 percent of the population, have laws mandating statewide e-waste recycling. Several more states are working toward passing new legislation and improving their existing policies. In several states, including California, Connecticut, Illinois, and Indiana, e-waste is banned from landfills. Check your state for legislation to have a better understanding of e-waste recycling laws. (www.ecycleclearinghouse.org/content.aspx?pageid=10).

Incinerators

Burning e-waste is the largest source of cancer-producing dioxins and also the largest source of heavy metal contamination in the United States Dioxins belong to the so-called dirty dozen, a group of dangerous chemicals known as persistent organic pollutants (POPs). Once dioxins enter the body, they last a long time because of their chemical stability and their ability to be absorbed by fatty tissue, where they are stored in the body, their half-life being estimated at 7 to 11 years. In the environment, dioxins tend to accumulate in the food chain. The higher an animal is in the food chain, the higher the concentration of dioxins. We are at the top of the food chain.

Incineration of electronic waste should be the last resort and should be at a minimum if not completely banned. Guiyu, China, which some call the e-waste capital of the world, receives 4,000 tons of the stuff per hour. It also has the highest-ever recorded level of dioxins, and 90 percent of its residents have neurological damage, an alarming rate of miscarriages in women, and the highest level of lead in children's blood in the world. Besides being stamped as a carcinogen, exposure to lead may be one of the most significant causes of violent crime in young people.

Dioxins and furans are two of the most toxic chemicals known today and when inhaled as gases, hydrochloric acid can form in the human lung causing ulcerations in respiratory tracts. Furthermore, dioxins and furans are regarded as very harmful, even at low doses, and can have serious health effects causing cancer, birth defects, and reduced immunity to other diseases, according to the EPA. Let's keep electronic waste as far away from landfills and incinerators as possible.

THE E-WASTE AFTER MARKET

People cannot be stopped from changing devices as often as desired, but they can be educated to discard old ones properly. The e-Stewards Initiative is an independent third-party certification program that assures responsible e-waste disposal by recyclers. In some cases, old electronics are reused, recertified, and resold, or sent to developing countries for reuse. However, in some cases, electronics sent to developing countries for reuse are only used a short time and then dumped in areas that don't have proper hazardous waste facilities.

There is a great deal of fake electronic recycling programs, which is why it's important to check out websites before recycling. E-waste can be a lucrative business because organizations can make a lot of money by exporting the items to developing countries that use the scrap metal rather than recycling them for reuse. There are several good resources to find local or regional e-waste recycling programs. The EPA and the Electronics Take Back Coalition offer information about e-Stewards. If there is no e-Steward near you, look at the manufacturer or store recycling programs. The Institute of Recycling Industries (ISRI) is the largest U.S. recycling industry association. The Coalition for American

Electronics Recycling (CAER) is another leading e-waste recycling industry association. The Electronic Products Recycling Association (EPRA) is the leading e-waste recycling industry association in Canada.

It's time that consumers take matters into their own hands and talk to manufacturers and their elected representatives in order to pass laws that will effectively increase the reuse of exotic, rare, and valuable material, to protect the health of the people and their environment—here and in the rest of the world.

SOURCES

"5 Types of Electronics You Need to Recycle." E-Terra Technologies, May 8, 2017, http://www.eterra.com.ng/news/5-types-electronics-need-recycle/.

"11 Facts about E-Waste." Do Something, https://www.dosomething.org/us/facts/11-facts-about-e-waste.

"17 Tips for E-waste Disposal at Home & Work." Olcese, November 1, 2016, https://olceseservices.com/2016/11/17-tips-e-waste-disposal-home-work/.

"Associated Industry-Led Electronics Recycling Programs in Canada." Electronic Products Recycling Association, http://epra.ca/who-we-are/related-links.

Ahmed, Faraz. "The Global Cost of Electronic Waste." *Atlantic*, September 29, 2016, https://www.theatlantic.com/technology/archive/2016/09/the-global-cost-of-electronic-waste/502019/.

Ala-Kurikka, Susanna. "Lifespan of Consumer Electronics Is Getting Shorter, Study Finds." *The Guardian*, March 3, 2015, https://www.theguardian.com/environment/2015/mar/03/lifespan-of-consumer-electronics-is-getting-shorter-study-finds.

"Battery Recycling: Battery Recycling Facts." University Library University of Illinois, http://guides.library.illinois.edu/battery-recycling.

Bradley, Laura. "E-Waste in Developing Countries Endangers Environment, Locals." *U.S. News and World Report*, August 1, 2014, https://www.usnews.com/news/articles/2014/08/01/e-waste-in-developing-countries-endangers-environment-locals.

Button, Kimberly. "20 Staggering E-Waste Facts." Earth 911, February 24, 2016, https://earth911.com/eco-tech/20-e-waste-facts/.

"Chemical Safety, Dioxins." World Health Organization, May 2010, http://www.wpro.who.int/mediacentre/factsheets/fs_201005_chemical_safety/en/.

Chen, Angus. "The Continent That Contributes the Most to E-Waste Is . . ." Goats and Soda, January 26, 2017, https://www.npr.org/sections/goatsandsoda/2017/01/26/511612133/the-continent-that-contributes-the-most-to-e-waste-is.

"Conquering the E-Waste Mountain." Computer Disposals Unlimited, May 14, 2015, http://www.computerdisposals.co.uk/news/conquering-ewaste-mountain/.

"Dioxins and Their Effects on Human Health." World Health Organization, October 4, 2016, http://www.who.int/en/news-room/fact-sheets/detail/dioxins-and-their-effects-on-human-health.

"E-Cycling: A Gentle Reminder of the Most Important Recycling 'Extra.'" Post Consumers, April 2, 2017, https://www.postconsumers.com/2017/04/02/e-cycling/.

Elliot, Matt. "What to Do with Your Old CD Collection." C/Net, April 22, 2014, https://www.cnet.com/how-to/what-to-do-with-your-old-cd-collection/.

"The Electronic Waste Crisis, Cell Phone Facts and Figures." Planet Green Recycle, November 1, 2015, https://planetgreenrecycle.com/fundraising/e-waste/the-electronic-waste-crisis-cell-phones-facts-and-figure.

"Electronic Revolution = E-Waste." The World Counts, Electronic Waste Facts, April 24, 2014, http://www.theworldcounts.com/stories/Electronic-Waste-Facts.

"E-Waste Crisis." Planet Recycle Green, https://planetgreenrecycle.com/fundraising/e-waste-problem.

"Facts and Figures on E-Waste and Recycling." Electronics Take Back Coalition, http://www.electronicstakeback.com/wp-content/uploads/Facts_and_Figures_on_EWaste_and_Recycling.pdf.

"Frequently Asked Questions." E-Stewards, http://e-stewards.org/faq/.

"Frequent Questions on Recycling." EPA, https://www.epa.gov/recycle/frequent-questions-recycling.

"Following the Trail of Toxic E-Waste." America's Toxic Electronic Waste as It Is Illegally Shipped to Become China's Dirty Secret, CBS 60 Minutes, November 6, 2018, https://www.cbsnews.com/news/following-the-trail-of-toxic-e-waste/.

Geyer, Roland. "The Economics of Cell Phone Reuse and Recycling." *International Journal of Advanced Manufacturing Technology*/ reprinted in *Research Gate*, March 2009, https://www.researchgate.net/publication/227231634_The_Economics_of_Cell_Phone_Reuse_and_Recycling.

"Global E-Waste: Statistics & Facts." Statista, https://www.statista.com/topics/3409/electronic-waste-worldwide/.

Haque, Tajirul. "Introduction to Electronics E-waste Recycling." *The Balance*, November 3, 2016, https://www.thebalancesmb.com/introduction-to-electronics-e-waste-recycling-4049386.

Harrington, Rebecca. "Here's What to Do with Your Dead Batteries." *Business Insider*, October 8, 2015, http://www.businessinsider.com/how-to-recycle-batteries-2015-10.

"Impacts of E-Waste on the Environment." E Terra Technologies, May 25, 2017, http://www.eterra.com.ng/articles/impacts-e-waste-environment/.

"ISRI Releases Study on the Economics of Recycling." Recycling Today, May 26, 2015, http://www.recyclingtoday.com/article/isri-2014-scrap-recycling-industry-economics/.

Lewis, Tanya. "World's E-Waste to Grow 33% by 2017, Says Global Report." *Live Science*, December 15, 2013, https://www.livescience.com/41967-world-e-waste-to-grow-33-percent-2017.html.

"Management of Electronic E-Waste in the United States, Analysis of Five Community Consumer Residential Collections of End-of-Life Electronic and Electrical Equipment." U.S. Environmental Protection Agency.

"Map of States with Legislation. "Electronic Recycling Coordination Clearinghouse, https://www.ecycleclearinghouse.org/contentpage.aspx?pageid=10.

"Mobile Devices: What Is the Average Lifespan of a Cell Phone?" *Quora*, https://www.quora.com/Mobile-Devices-What-is-the-average-lifespan-of-a-cell-phone.

"Printer Cartridge Recycling." Earth Share, http://www.earthshare.org/2010/09/green-quiz-answer-printer-cartridge-recycling.html.

Noyes, Katherine. "Can 'Urban Mining' Solve the World's E-Waste Problem?" *Fortune*, June 26, 2014, http://fortune.com/2014/06/26/blueoak-urban-mining-ewaste/.

Office of the Inspector General. *A Review of Federal Prison Industries' Electronic-Waste Recycling Program*. Washington, DC: U.S. Department of Justice, 2010, https://oig.justice.gov/reports/BOP/o1010.pdf.

Recycling Ink Cartridges, Canon, https://www.usa.canon.com/internet/portal/us/home/support/self-help-center/recycling-ink-cartridges.

"Release of Dioxins from Solid Waste Burning and Its Impacts on Urban Human Population—A Review." *Journal of Burning Effects & Control*, https://www.omicsonline.org/open-access/release-of-dioxins-from-solid-waste-burning-and-its-impacts-on-urbanhuman-population-a-review-2375-4397-1000215.pdf.

Risen, Tom. "America's Toxic Electronic Waste Trade." *U.S. News and World Report*, April 22, 2016, https://www.usnews.com/news/articles/2016-04-22/the-rising-cost-of-recycling-not-exporting-electronic-waste.

Scrogum, Joy. "Electronics Take Back Coalition ETBC, Best Buy Ends Free Recycling of Televisions and Monitors." Sustainable Electronics Initiative, February 12, 2016, http://wp.istc.illinois.edu/sei/category/electronics-takeback-coalition-etbc/.

Shore, April. "Dioxins and Their Effects on Human Health." *Blogspot*, World Health Organization, October 4, 2016, http://www.who.int/en/news-room/fact-sheets/detail/dioxins-and-their-effects-on-human-health.

"State Legislation." Electronics Takes Back Coalition, http://www.electronicstakeback.com/promote-good-laws/state-legislation/.

Statt, Nick. "Why Apple and Other Tech Companies Are Fighting to Keep Devices Hard to Repair." *The Verge,* August 3, 2017, https://www.theverge.com/2017/8/3/16087628/apple-e-waste-environmental-standards-ieee-right-to-repair.

"The Sad Truth about E-Waste: Top 10 Things You Need to Know." *e-Cycle Florida*, http://www.ecycleflorida.com/ecycle-blog/1797/.

"The e-Stewards Story." e-Stewards, http://e-stewards.org/about-us/the-e-stewards-story/.

Vargas, June. "Don't Waste Your E-Waste: Put It to Work." State of California Department of Consumer Affairs, December 29, 2016, https://thedcapage.blog/2016/12/29/dont-waste-your-e-waste-put-it-to-work/.

"What Is E-Waste Recycling?" *Conserve Energy Future*, https://www.conserve-energy-future.com/e-waste-recycling-process.php.

Zhoa, Bin, Minghui Zheng, and Guibin Jiang. "Dioxin Emissions and Human Exposure in China: A Brief History of Policy and Research." *Environmental Health Perspectives*, March 2011, https://www.ncbi.nlm.nih.gov/pmc/articles/PMC3060009/.

13

OPENING PANDORA'S PHARMACY

> If all the drugs were thrown in the ocean, everyone would be better off . . . except the fish. —*Oliver Wendell Holmes*

There is a persistent and pervasive, high-priced illness in America today. It's the exceedingly expensive, overly prescribed, and incredibly wasted amounts in American medicine, under and over the counter (OTC), in concert with doctors and America's big pharma. It's hand in hand in glove with one of the worst effects of modern medical technology—the belief that whatever ails us, take a pill to kill it no matter what the cost or waste.

One of the biggest increases in American medicine in the last half-century is the increase in the number of medications prescribed to patients, partly due to technology, partly due to the business of selling sickness, and also the jump to prescribe. The average American takes about 12 medications annually compared to seven 20 years ago. Back then, spending on drugs totaled about 5 percent of the total U.S. health care costs, now it's more like 17 percent. Spending on prescription medications has increased by a knockout of $200 billion in two decades.

Incidentally, by most estimates, the United States spends between $3 and $4 trillion on health care annually. According to a 2015 Bloomberg report, the United States. is the third most expensive country for medical care, yet ranked 50th in efficiency.

Americans are older and heavier, with more hypertension, high cholesterol, diabetes, osteoporosis, and arthritis than a generation ago—all conditions thankfully and effectively treated with medications. In 2016, there were approximately 4.45 billion prescriptions issued in the United States. That averages to something like 13.7 prescriptions filled for each person in the country, according to Statista.com. Approximately 55 percent of Americans take an average of four prescription drugs, and 75 percent also take at least one or more OTC drugs regularly.

The average annual rate for prescriptions is now almost four prescriptions per child (from birth to age 18), and according to *Forbes* magazine and Medscape.com the senior population, age 65 to 79, comprises 13.7 percent of the population but uses 40 percent of all prescription drugs. People over 70 fill an average of 27 prescriptions for new drugs per year.

We take a total of 2.9 billion trips annually to purchase retail OTC products. On average, U.S. households spend about $338 per year on OTC products, tens of billions on vitamins, supplements, diet pills, cold cures, herbal remedies, and other medicaments— incredible economic benefits to the companies that manufacture them.

Part of the growth of the almost half-trillion-dollar medical industry has been brought about by waste—both in overprescribing by the pharmaceutical and medical industry and in the discarding of medicine by consumers. Lamentably, the ways much of the drug waste is being disposed of is causing a major health hazard. Consumers aren't the only ones tossing their drugs down the drain or in the garbage. The Associated Press estimates that hospitals and long-term medical care institutions across the United States are dumping 250 million pounds of pharmacologically active drugs directly into public sewer systems each year that can have severe effects on humans and wildlife.

TOSSING MEDS MEANS TOXIC WASTE

Disturbing studies of waste water near hospitals in Europe and the United States have found high concentrations of antibiotic-resistant bacteria and of organisms with genetic mutations similar to those that can cause cancer in humans and could cause changes to bacterial DNA. Many pharmaceutical drugs, such as painkillers, antidepressants, and hormones, are now being detected in creeks and bays, including an alarming amount of acetaminophen in the San Francisco Bay.

Don't flush your unwanted medicine down the toilet or put it in the garbage. Studies have shown that chemicals from pharmaceuticals are inhibiting reproduction in fish found in many lakes and rivers. Dispose of unwanted or expired pharmaceuticals at a household hazardous waste facility or a pharmacy that accepts medication for proper disposal.

Mishandled Meds

In 2010, an estimated 5.9 million tons of medical waste was produced in hospitals, much of it dangerous and/or hazardous. Discarded needles may expose waste workers to potential puncture injuries and infection when containers break open inside garbage trucks or needles are mistakenly sent to recycling facilities and poke through plastic garbage bags, or found as waste on the beach.

With Environmental Protection Agency's (EPA's) tightly controlled Hospital, Medical, and Infectious Waste Incinerators (HMIWI) standards, the number of incinerators in the United States has declined since 1997. This has led to an increase in the use of alternative technologies for treating medical waste, rendering it noninfectious and enabling it to be disposed of as solid waste in landfills or incinerators.

According to Express Scripts (a mail-order pharmaceutical company):

- $55.8 billion was wasted on higher-priced medications when similar, affordable drugs could have been used.
- $93.1 billion could have been saved if patients had shopped around or used mail-order pharmacies.
- $269.4 billion was spent on avoidable medical expenses because patients did not stay on the medicines they were prescribed.

A 2012 *Journal of the American Medical Association* (*JAMA*) paper, identified six major areas of approximately $765 billion annually in waste that included unnecessary treatments, overpriced drugs and procedures, and the underuse of care in disposal, making up a whopping 34 percent of the total U.S. health care spending including:

- Unnecessary services, $210 billion.
- Inefficient delivery of care, $130 billion.

- Excess administrative costs, $190 billion.
- Inflated prices, $105 billion.
- Prevention failures $55 billion.
- Fraud, $75 billion.

Surplus for the World's Sick

Some of this waste has been turned into recovered resources by Partners for World Health (PWH), a nonprofit run by a registered nurse, Elizabeth McLellan, which ships surplus unused equipment overseas to developing nations. Among their 15,000-square-foot warehouse shelves there are unopened packages of everyday medical equipment that would be sent to landfills if not for her project. Last year, PWH sent seven containers overseas, each weighing up to 15,000 pounds with an estimated value of $250,000. And this is not the only nonprofit organization doing this work.

One of the bags in a 10-foot mound of boxes and bulging sacks at the PWH warehouse contains unopened and unexpired hospital equipment, from sterile bandages to a pulse oximeter adapter (used to monitor oxygen saturation of a patient's blood). "This is normal for every hospital," McLellan says. The "discards" in the PWH warehouses that would otherwise end up dumped are worth approximately $20 million. To make matters worse, there are more than 600 rural hospitals in the United States that are so strapped financially they risk closure.

Researchers at the University of California San Francisco Medical Center, estimate that in a single year, the hospital wasted $2.9 million in neurosurgery supplies—that's in one department alone. Dr. Corinna Zygourakis, the chief neurosurgery resident at the San Francisco Medical Center, and her colleagues conducted a study to see if that waste could be reduced. Surgical teams were given an incentive of a bonus for their department if they could reduce their medical equipment costs by at least 5 percent. Her team's supply costs dropped by 6.5 percent during the study period, while a control group's costs increased by almost 7.5 percent.

Out of Date, But Not Obsolete

The U.S. Food and Drug Administration (FDA) study tested more than 100 medications that had been stockpiled by the military, and found that 90 percent of them were safe and effective for years after the expiration date, yet they were dumped.

The problem is that drug manufacturers don't study what happens to many drugs over long periods of time. For that reason, a conservative, but responsible approach is recommended. Replace the meds a year after getting them. If you do choose to take an expired pill, the main risk you're taking is that it won't work. For a headache remedy or cold medicine, it's worth the risk, but for a medication that's crucial for your health, you want to make sure it's going to do its job. Most meds, with the exception of tetracycline, don't become toxic or dangerous after they expire. Liquids and liquid capsules tend to be less stable than solid tablets, and storing medications properly can help extend their life. By the way, a cabinet in a humid bathroom is probably the worst place to keep drugs. A cool, dry place is far better.

Medication Mistakes—A Multi-billion Boondoggle

The United States wastes $418 billion on what is called "bad medication-related decisions." What's more, in a result of findings by the pharm giant CVS, the waste was worst in states that had the lowest incomes.

Perhaps by far the biggest waste has to do with how medicine is practiced—unnecessary services; overprescribed meds, inefficient care, or the failure to prevent problems that require less expensive intervention, and medical mistakes.

Perfectly safe, up-to-date medications—already paid for, often by federal or state governments—are being discarded by more than 16,000 nursing homes and other long-term care facilities around the country. Meds that could be redistributed are being thrown out, while one in every four people in the United States struggle to afford their prescriptions. About $700 million worth of medications could be salvaged each year in 10 million prescriptions.

The money lost or spent on medical waste in the United States is displacing spending on critical national infrastructure needs.. It's been estimated that this annual waste could have paid for the insurance coverage of both the employer and employee contributions of 150 million workers. In addition, the industry spends well in excess of $21 billion a year promoting their products, some of which are already overpriced and overprescribed.

A PILL FOR EVERY ILL

Many of us and our physicians still think that every symptom, every hint of disease requires a drug. A recent comprehensive review of studies cited in Pubmed.com of such inappropriate prescribing in older patients found that 21.3 percent of patients 65 years or older were using at least one inappropriately prescribed medication.

The amount of harm stemming from inappropriate prescriptions is staggering. More than 1.5 million people are hospitalized and more than 100,000 die each year from largely preventable adverse reactions to drugs that should have been prevented. Almost 1.3 million people went to emergency rooms due to adverse drug effects in 2014, and about 124,000 died from those events, according to estimates based on the Centers for Disease Control and Prevention (CDC) and the FDA.

Before you leave a doctor's office, don't be afraid to ask why each drug is being prescribed and what it should be doing. Also, do not hesitate to ask if there are alternative treatments that are less expensive or don't involve a prescription.

Vitamin Folly

Vitamins are actually a collection of micronutrients that are essential in fairly minute quantities for your body to functionally normally. We usually get plenty in daily diets, and if you take more than the suggested dosage, they will be eliminated from you (down the toilet) and from your wallet.

Vitamin supplement companies would have you believe that vitamins are the cure for the common cold, are the best way to remain healthy, prolong life, and ensure and reinforce a good diet—all are myths. About half of Americans take multivitamins, many also take individual vitamin supplements. And roughly one in five adults uses a herbal supplement. Stores reinforce the idea that these are needed, as the sales of pills, capsules, and powders make up 5 percent of all grocery sales in the United States with profit margins about 10 times as high as those of food items. In fact, some consumers think that if a supplement provides 100 percent of their needs, then something that provides more must be better. The truth is, it doesn't work that way.

In fact, in some cases this can be harmful—especially in large doses of vitamins C, D, and E. High doses of vitamin C are known to cause gastric discomfort and diarrhea, and the current upper tolerable daily intake is 2,000 mg for adults daily. It's always safer to take less than the daily dose determined by the manufacturer. Vitamin E in large doses can be harmful to blood clotting, and could cause prostate cancer in men.

"Fifty percent of Americans are at suboptimal levels for vitamin D," says Donald Hensrud, MD, director of the Mayo Clinic Healthy Living Program. "But be careful about taking Omega 3 for vitamin D because that could build up and make you sick," he cautions. "Eat fish and use olive oil instead."

Thirty percent of millennial women take a multivitamin. But research on supplements shows that although a few might make you feel healthier, especially psychologically, many are a waste of money or, worse, actually dangerous. The FDA doesn't limit the dosages or evaluate the safety of these ingredients.

The cost of the vitamins doesn't mean better or healthier. Plant-based vitamins from a pricey health-food store can cost three times as much as standard drugstore synthetics, but they're the same thing. Vitamin C from rose hips isn't better than synthetic C, and neither is calcium from oyster shells. Just because supplements are "organic" doesn't mean they can't be contaminated if the soil, plants, or water where they were harvested was impaired.

Some multivitamins add herbs, but they don't tell you how much you're getting. Herbs can be high in lead, arsenic, or cadmium. If you want to take an herb, take a single one on its own, which has a lower chance of being contaminated. Forget the Flintstone's brand. It's difficult for manufacturers to control the amount of ingredients like sugar or substitutes in gum balls and mints, and it's easy to go overboard with them because they taste good.

Meth Heads and Head Colds

Years ago, the FDA removed pseudoephedrine, a nasal decongestant in Sudafed, Claritin-D, and other allergy aids from the shelves—not because it didn't work or was dangerous, but because meth heads were grinding them up to brew crank. So out went pseudoephedrine, in came phenylephrine, an FDA-approved ingredient.

In a new study of more than 500 adult allergy sufferers researchers found that phenylephrine, was no better at curing their ailment than a placebo—even when given at higher doses than those currently approved. The study called on the FDA to pull its approval of phenylephrine from its list of effective nasal decongestants. Other researchers from the University of Florida stated that people have been throwing money away in the name of nasal relief, meaning that the $8 billion the drug companies rake in annually selling OTC cold medication is basically "down the toilet" for your money and your sinuses.

The National Health Service (NHS) says there's not much scientific evidence that cough medicines work and there are no shortcuts with coughs caused by viral infections, it just takes time for your body to it fight off. Some may contain dextromethorphan (DM), it's something to ease your aching head, dry up streaming noses and eyes, and this may be able to help get you to sleep, but it won't kill the bug, just mask the symptoms. Keep in mind that a cough is part of the natural healing, productive process.

Four widely used cold remedies are:

- Zicam. It's website makes the misleading claim that products are regulated by the FDA. They're not.
- Airborne. It doesn't cure anything. It's a cleverly marketed vitamin supplement with no scientific support for any health benefits and had to pay $23 million back in 2008 to settle a class-action lawsuit over its advertising.
- Coldcalm not only doesn't work, but it contains belladonna (a natural poison). Another ingredient is pulsatilla, which is highly toxic, and produces cardiogenic toxins and oxytocin that slow the heart in humans.
- Umcka's active ingredient is a homeopathic plant extract called pelargonium sidoides, an African geranium that really doesn't do much.

The things that you can't get at the drug store like mom providing warm humid air, heated honey and tea, keeping you hydrated, tender care, and, yes, chicken soup, work as well as anything, and cost a heckeva lot less than what's on the shelves of your pharmacy.

Home Brew Herbs

About one in five adults uses herbal supplements, and altogether they will spend $21 billion on vitamins and herbal supplements each year. Incidentally, a 2015 New York Attorney General investigation and lab tests that focused on a variety of herbal supplements from four major retailers: GNC, Target, Walmart, and Walgreens, determined that only 21 percent of the products actually had DNA from the plants advertised on the labels. Another test, reported in the *New York Times*, stated that one-third of the 44 herbal supplements tested showed outright substitution, meaning there was no trace of the plant advertised on the bottle—truly a waste of money, if not also criminal.

There are shelves full of homeopathic remedies from echinacea, brewer's yeast, and Sambucol, to various teas, goldenseal, and exotic herbs. These remedies claim to fight a hacking, tickling cough, a head cold, body aches, and what have you. But according to studies by the National Institutes of Health (NIH), there's little to no hard evidence that shows they work effectively, and are most likely another waste of your money. In fact, these products mainly work through positive thinking and mind over illness.

Be careful of combining herbal supplements with your regular meds. For example, oral contraception for birth control can be rendered inactive if taken with the herbal supplement, St. John's Wort, which can also interfere with the effectiveness of other medications, including antidepressants.

Prices for the various OTC pain killers, vitamins, and other medicines can fluctuate wildly from basic aspirin for a couple of bucks a bottle to $30 for a specialty brand name. Choose wisely; just because it has been "branded" doesn't mean it is more effective. In fact, many of the chemical makeups for generics are basically the same for a brand name.

DON'T-WORK DIET PILLS

The idea of simply taking a pill to shed extra weight is hard to resist. Americans now spend more than $30 billion per year on dietary supplements. We are bombarded with quick weight-loss ads on late-night TV, especially just before tank top, shorts, and swimsuit season. The words "quick weight loss" on a search engine will cough up millions of hits in the dietary supplements business. Diet pills generally fall into one of three categories. They are either powerful laxatives, stimulants, or appetite suppressants.

A *Consumer Reports* survey of almost 3,000 Americans found that nearly one in four people take weight-loss supplements, and roughly 20 percent erroneously believed that the FDA guarantees their safety and effectiveness. More than half of the people surveyed said they had uncomfortable side effects while taking a weight-loss supplement such as rapid heart rate, dry mouth, and digestive issues. And diet supplements can interact with the prescription drugs you're already taking, which can result in some serious problems. In fact, before intermingling any kind of supplement, herb, or drug, ask your doctor.

Pills, nonprescription drugs, herbal medicines, or other dietary supplements are at best gimmicks that may help some with some weight loss, for a while. There is relatively little research about the efficacy of these products and some have proven dangerous, even deadly. Remember Fen-phen—it's back in disguise. Fenfluramine was one of the ingredients in "Fen-phen," and dexfenfluramine is closely related to fenfluramine, so it's basically the same thing just renamed and repackaged.

If the bottle reads, "natural," or "extreme fat burner" and makes claims that you'll "crave less" or it will "reset your metabolism," and it will do this in weeks, don't believe it. There's no magic in those magic bullets.

According to the website Compounding Pharmacy of America here are some of the most dangerous diet pills, which have been banned or have received serious warnings for use as diet supplements by the FDA.

- Meridia
- Clenbuterol
- Ephedrine
- Dinitrophenol

- Fen-phen
- Japan Rapid Weight Loss Diet Pill
- The Brazilian Diet Pill
- Contrave
- Qnexa

The naked truth is that the most effective way to lose weight and keep it off is through lifestyle changes that are simple but tough: eating a healthy low-calorie, low carbohydrate diet, lots of fruits and vegetables, and physical activity. Taking any medication, especially by prescription, is between you and your physician and should be monitored by patient and doctor.

Even products that aren't necessarily harmful may simply be a waste of money. For example, Hoodia, a cactus-like plant from Africa, is marketed as an appetite suppressant despite the fact that there are no studies of its safety or effectiveness. It's also unclear how much real Hoodia is in these products since harvesting is limited by conservation laws.

Research—Don't Try before You Buy

OTC drugs are considered supplements and they do not require FDA approval and there is most likely no hard evidence to support their claims. It's simply not possible for the FDA to keep up with the diet or supplement pill industry because it's just too large. Many contain ingredients not even listed on the label and some supplements have been shown to increase heart rate or blood pressure, which can lead to heart attack or stroke risks, and there have been reports of people having seizures after using unregulated diet substances.

It's important to do your homework about OTC weight-loss pills. General information about many dietary supplements is available at the website of the National Center for Complementary and Alternative Medicines. The database summarizes research regarding dietary supplements and herbal products. Although the database is only available by subscription, you may be able to access it through a public library. The FDA has more detailed information about the risks of weight-loss products on its website (https://www.fda.gov). If you want to report problems yourself, contact the FDA at 800-FDA-1088 or online. And check out the NIH's Office of Dietary Supplements for research briefs to learn which supplement claims are really worth listening to.

MED MANUFACTURER'S MANEUVERS

Patients, insurance companies, and the federal government are on track to spend about $3 billion this year on cancer drugs that will be thrown away, according to a report in the *BMJ* (formerly the *British Medical Journal*), and CBS News. The drugmakers, hospitals, and doctors have been profiting prolifically from the way these drugs are packaged. Researchers at Memorial Sloan Kettering Cancer Center and the University of Chicago determined that one-third or more of some drugs are not used because prepackaged doses are too large and safety regulations require them to be thrown away.

Here's a for instance. Keytruda, a new cancer drug, comes in a 100-mg vial and costs approximately $5,000. The manufacturer, Merck, packages this drug only in a 100-mg size. However, a typical patient needs 150 mg, which means half of the second vial is actually thrown away—or $2,500 worth of the pricey valuable drug is wasted. Merck does claim that it's moving to a fixed-dose 200-mg size that will eliminate waste, but only after complaints, and making millions on the medicine, much of it wasted.

Although doctors and hospitals sometimes use leftover drugs to treat a subsequent patient, this practice is very limited, partly due to safety standards from the U.S. Pharmacopeial Convention that permit sharing only if the leftover drug is used within six hours.

A study, published in the *BMJ* concluded that Medicare and private insurers, as well as patients, pay companies about $1.8 billion a year for medications that are thrown away. That's 10 percent of drug companies' projected 2016 revenue of $18 billion. Add another $1 billion to doctors and hospitals as price markups on those discarded medications. Those excess profits would not exist if companies sold vial sizes more in line with the needs of patients, researchers suggest, and the problem is not unique to cancer drugs.

HOPE OR HYPE FOR HELP WITH AGING

You've seen the ads, the adoring looks for the twelve-pack on the glistening abs of men, the tight gluts of women pumping on bikes, the not-so-subtle hints from guys with white-wall hair desiring to give their energy, libido, and body a tune-up. The message is delivered from retired athletes, or frustrated middle-aged moms, but the message is the same—buy this remedy and get some of your youth, looks, sex, and energy back, now, before it's too late. Ponce de Leon's quest for the fountain of youth has been picked up by the promise of the supplement industry today—and both are driven mainly by hype, hope, and hoax.

The lure of OTC antiaging pills and potions can be easy to swallow, after all, it's human nature to want to look and feel younger, especially by simply taking a nostrum. We know how to slow the aging process: healthy living, a good diet, plenty of exercise, no smoking, moderate use of alcohol, is probably the best bet—but there are no guarantees. After all, Jim Fixx, a marathon runner in supposedly terrific health stroked out on a jog. Aging is a complex process not to be eluded by an elixir; there's no cure for the inevitable.

But "antiaging" advocates insist that there are supplements that not only prevent or delay effects of aging, but can give you back some of the verve and vigor that living can exhaust. Popular antiaging supplements supposedly include hormones that decrease as we age. Proponents say that taking hormones, antioxidants, and amino acids can control, prevent, or even reverse age-related problems, such as a decrease in cardiovascular function, muscle mass, strength, sexual appetite, as well as the detrimental effects of "free radicals." They're extremely reactive atoms or molecule fragments that steal electrons from other molecules in body tissues in a process called oxidation, damaging cells and speeding up the aging process.

Touted as the "super hormone" or even "nature's antidote to aging," DHEA (dehydroepiandrosterone) has attracted the attention of gerontologists and is widely promoted in antiaging programs and clinics. It's a precursor to testosterone and estrogen, the hormones that gives men their masculine traits, while estrogen is the hormone that makes women, well, women.

A study by the Baltimore Longitudinal Study of Aging (BLSA) has been carefully examining the aging process in more than 1,000 people from ages 20 to 90. DHEA has been extensively studied—in test tubes, lab animals, men, and women. Overall, the human research does not support the antiaging claims. One of the longest studies on DHEA was a well-designed two-year Mayo Clinic trial of people over 60 in the *New England Journal of Medicine* in 2006. It found that DHEA supplements did not improve muscle strength, sexual function, athletic or physical performance, body and bone composition, blood sugar control, cognitive function, or general quality of life.

And its potential risks are numerous. There are very few reasons to take it save for vanity, and probably misguided hope. But because it is not regulated by the FDA, the quality or quantity of the drug is probably unknown. Besides, most of the DHEA samples studied are not from mammals, but from a plant source called diosgenin, which is converted to DHEA in a lab.

But if you want to heap cash on the problem, try the Ambrosia solution. The company has jumped on the "young blood" phenomenon and is currently offering transfusions of plasma taken from young people, reportedly at a cost of $8,000 per pop, claiming that it will cause a rejuvenation in older people. As yet, clinical trials have produced nothing more than a demonstration that old mice increased neuron growth and improved memory after about 10 infusions of blood from young mice.

Another age-defying supplement that has been getting a barrage of late-night PR is a drug called Prevagen. The claim is that it will slow or aid in the loss of memory, concentration, and forgetfulness.

Apparently, the FDA is trying to shut down the claims as they did with another brain supplement, Luminosity. Their claim was that there is no way for Prevagen to work, as the "active ingredient" is a protein, apoaequorin. The GI system breaks down all proteins into amino acids, the building blocks of all proteins. So, it ends up the same as eating a veggie burger.

The arguments between the government and the manufacturer of Prevagen, Quincy Bioscience, grew heated enough for the Federal Trade Commission and New York State Attorney General to charge the marketers of Prevagen of making false claims. According to the New York Attorney General, "The marketing for Prevagen is a clear-cut fraud, from the label on the bottle to the ads airing across the country. It's particularly unacceptable that this company has targeted vulnerable citizens like seniors in its advertising that provides none of the health benefits that it claims."

Quincy Bioscience replies, "We vehemently disagree with these allegations made by only two FTC commissioners. This case is another example of government overreach by imposing arbitrary new rules on small businesses like ours," a "small" multimillion-dollar business, by the way. The lawsuit against Quincy Bioscience was dismissed by a federal judge in 2017. The court rejected arguments by the FTC and New York Attorney General and ruled that the government had not shown that the company's claims were wrong, just that there was an increased risk that they might be wrong.

What you can do to help put a halt to this craziness and waste is to watch your own lifestyle and be positive and sensible about how you take care of your mind and body—for now that's the only real fountain of youth.

SOURCES

A., Karen. "Are We Over Medicated? Do We Use Pills Instead of Living Healthy?" *MeetUp/Huffington Post*, October 2012, https://www.meetup.com/tampa-bay-thinkers/events/246797304/.

Abbott, Alden. "FDA Reform: A Prescription for More and Better Drugs and Medical Devices." The Heritage Foundation, June 20, 2016, https://www.heritage.org/government-regulation/report/fda-reform-prescription-more-and-better-drugs-and-medical-devices.

Albaum, Shelly. "Why I Feel Suckered by Elysium Health." Right of Assembly, August 16, 2017, https://www.right-of-assembly.org/single-post/2017/08/16/Why-I-Feel-Suckered-by-Elysium-Health.

Allen, Marshall. "The Staggering Cost of Medical Waste in America." The Fiscal Times/ProPublica, March 10, 2017, http://www.thefiscaltimes.com/2017/03/10/Staggering-Cost-Medical-Waste-America.

"Americans Spend Billions on Vitamins and Herbs That Don't Work." Health Lines, https://www.healthline.com/health-news/americans-spend-billions-on-vitamins-and-herbs-that-dont-work-031915#1.

"Amino Acids and the Fight against Aging." AminoAcidsStudy.org, http://aminoacidstudies.org/anti-aging/.

Angell, Marcia. "The Truth about the Drug Companies." *New York Review of Books*, July 15, 2004, http://www.nybooks.com/articles/2004/07/15/the-truth-about-the-drug-companies/.

Ball, Philip. "A Bitter Pill to Swallow: Cough Medicines Won't Cure You." *Guardian*, October 2012, https://www.theguardian.com/commentisfree/2012/oct/16/cough-medicine-wont-cure-you.

Bartol, Tom. "Which Drugs Should Be Deprescribed in the Elderly?" *Medscape*, April 26, 2018, https://www.medscape.com/viewarticle/847187_2.

Berwick, Donald, and Andrew Hackbarth. "Eliminating Waste in U.S. Health Care." *JAMA*, April 11, 2012, https://jamanetwork.com/journals/jama/article-abstract/1148376?redirect=true.

Dandurant, Karen. "Are Diet Pills a Waste of Money?" Edge Radio, April 26, 2015, http://www.fosters.com/article/20150426/NEWS/150429490.
"Diet Pills You Should Avoid." Compounding Pharmacy of America, https://compoundingrxusa.com/blog/dangerous-diet-pills-avoid/.
"Don't Flush Your Medicines Down the Toilet!" http://pharmacy.ca.gov/publications/dont_flush_meds.pdf.
"Eat a Balanced Diet." Healthline, https://www.healthline.com/health/balanced-diet.
Feller, Stephen. "$3 billion in Cancer Drugs Wasted Every Year Due to Packaging." UPI, March 1, 2016, https://www.upi.com/Health_News/2016/03/01/3-billion-in-cancer-drugs-wasted-every-year-due-to-packaging/3231456845488/.
"FDA Regulation of Drugs versus Dietary Supplements." American Cancer Society, https://www.cancer.org/treatment/treatments-and-side-effects/complementary-and-alternative-medicine/dietary-supplements/fda-regulations.html.
Gorenstein, D. "Bad Packaging Means $3b in Drugs Will Be Wasted This Year." WBFO News, March 1, 2016, http://news.wbfo.org/post/bad-packaging-means-3b-drugs-will-be-wasted-year.
Goldman, Chelena. "This Is the Average Number of Rx Meds Americans Take Daily." CheatSheet, November 21, 2017, https://www.cheatsheet.com/health-fitness/how-many-rx-meds-does-the-average-american-take.html/?a=viewall.
Gorenstein, Dan. "The Shocking Cost of Wasted Prescription Pills." Marketplace, December 10, 2014, https://www.marketplace.org/2014/12/10/health-care/shocking-cost-wasted-prescription-pills.
"Government Lawsuit Disputes 'Extravagant' Claims of Memory Supplement." CBS News, January 10, 2017, https://www.cbsnews.com/news/prevagen-lawsuit-supplement-memory-improvement-claim-quincy-bioscience/.
Gross, Jane. "James Fixx Dies Jogging, Author Was 52." *New York Times*, July 22, 1984, https://www.nytimes.com/1984/07/22/obituaries/james-f-fixx-dies-jogging-author-on-running-was-52.html.
Gutierrez, David. "Hospitals Flush 250 Million Pounds of Expired Drugs into Public Sewers Every Year." *Natural News*, February 10, 2009, https://www.naturalnews.com/025573.html.
Heckler, Raquel. "5 Ways Vitamin D Could Save Your Life." ABC News, August 3, 2009, http://abcnews.go.com/GMA/Parenting/story?id=8234947&page=1.
"Herbal Supplements Filled with Fake Ingredients, Investigators Find." CBS News, February 3, 2015, https://www.cbsnews.com/news/herbal-supplements-targeted-by-new-york-attorney-general/.
Herper, Matthew. "Does Medication Waste Cost the U.S. $418 Billion?" *Forbes*, June 28, 2013, https://www.forbes.com/sites/matthewherper/2013/06/28/does-medication-waste-cost-the-u-s-418-billion/#d7208e825e96.
"How to Lose Weight and Keep It Off." HelpGuidce.org, https://www.helpguide.org/articles/diets/how-to-lose-weight-and-keep-it-off.htm.
"Hospitals Waste Millions on Discarded Unused Medical Supplies." Health Care Finance. September 12, 2016, http://www.healthcarefinancenews.com/news/hospitals-waste-millions-discarded-unused-medical-supplies.
Jaffee, Lynn. "The Nine Biggest Weight Loss Mistakes," Acupuncture Health Insights, March 18, 2011, http://acupuncturetwincities.com/2011/03/the-nine-biggest-weight-loss-mistakes/.
Kantrowitz, Barbara. "Weight Loss: How to Spot Dangerous Diet Pills." *Newsweek*, March 31, 2009, http://www.newsweek.com/weight-loss-how-spot-dangerous-diet-pills-75887.
Levine, Beth. "The Truth about Generic vs. Brand-Name Medications." *Huffington Post*, February 22, 2015, https://www.huffingtonpost.com/2015/02/22/generic-prescriptions_n_6730194.html.
Los Angeles 2015 Healthy Aging Report. University of Southern California, 2005, http://roybal.usc.edu/wp-content/uploads/2016/04/USC_Roybal-LA_HealthyAging.pdf.
Marshall, Allen. "What Hospitals Waste." *ProPublica*, March 9, 2017, https://www.propublica.org/article/what-hospitals-waste.

Mangan, P. D. "Anti-Aging Technology Hype and Reality." *Rogue Health and Fitness*, July 3, 2017, http://roguehealthandfitness.com/anti-aging-technology-hype-reality/.

"Medical Waste: Why American Health Care Is So Expensive." Wharton, University of Pennsylvania, August 18, 2016, http://knowledge.wharton.upenn.edu/article/medical-waste-american-health-care-expensive/.

"Medical Waste." U.S. Environmental Protection Agency, https://www.epa.gov/rcra/medical-waste.

Mehndiratta, Detriment. "Over-the-Counter Drug—Need to Know Its Benefits." Health and Fitness, March 23, 2018, http://ezinearticles.com/?Over-The-Counter-Drug---Need-to-Know-Its-Benefits-and-Detriment&id=9908987.

Mole, Beth. "Common Decongestant May Be Worthless, Study Finds." Ars Technical, October 29, 2015, https://arstechnica.com/science/2015/10/common-decongestant-may-be-worthless-study-finds/?comments=1.

"My Wife Claims That Sudafed 24 Hour Helps Her Allergies, but Claritin 24 Does Not!" All Nurses, http://allnurses.com/general-nursing-discussion/my-wife-claims-79604.html.

Nicks, Denver. "How Big Pharma Makes Billions on Cancer Drugs That Are Thrown Away." *Money*, March 1, 2016, http://time.com/money/4243027/drug-packaging-waste/.

Noonan, Michele. "Herbs That Affect Birth Control Pills." *Live Strong*, August 14, 2017, https://www.livestrong.com/article/375674-herbs-that-affect-birth-control-pills.

"Offer Your Patients Relief from Allergies." Ancillary Medical Solutions, http://www.ancillarymedsolutions.com/allergy.php.

"Overspending Driven by Oversized Single Dose Vials of Cancer Drugs." *BMJ*, March 2016, https://www.bmj.com/content/352/bmj.i788.

"Pollution Facts." Save the Bay, https://www.savesfbay.org/pollution-facts.

"Prevagen: How Can This Memory Supplement Flunk Its One Trial and Still Be Advertised as Effective?" Center for Science in the Public Interest, November 20, 2017, https://cspinet.org/news/prevagen-how-can-memory-supplement-flunk-its-one-trial-and-still-be-advertised-effective.

"Prescription Drugs." Georgetown University, https://hpi.georgetown.edu/agingsociety/pubhtml/rxdrugs/rxdrugs.html.

"Prescriptions per Capita in the U.S. in 2013, by Age Group." Statista, https://www.statista.com/statistics/315476/prescriptions-in-us-per-capita-by-age-group/.

"Progesterone." Revolvy, https://www.revolvy.com/main/index.php?s=Progesterone&sr=50.

Rastogi, Nina. "Wasting Syndrome: How Much Trash Do Hospitals Produce?" *Slate*, October. 19, 2010, http://www.slate.com/articles/health_and_science/the_green_lantern/2010/10/wasting_syndrome.html.

"Retail Prescription Drugs Filled at Pharmacies Annual per Capita by Age." Henry J. Kaiser Foundation, https://www.kff.org/other/state-indicator/retail-rx-drugs-by-age/?currentTimeframe=0&sortModel=%7B%22colId%22:%22Location%22,%22sort%22:%22asc%22%7D.

Rosick, Edward R. "DHEA: The Hormone of Youth?" Life Enhancement, 2018, http://www.life-enhancement.com/magazine/article/776-dhea---the-hormone-of-youth.

Salzberg, Steven. "The Top Five Cold Remedies That Do Not Work." *Forbes*, November 17, 2014, https://www.forbes.com/sites/stevensalzberg/2014/11/17/the-top-five-cold-remedies-that-do-not-work/#fcfbc774701b.

Scott, Cameron. "Americans Spend Billions on Vitamins and Herbs That Don't Work." Healthline, March 19, 2015, https://www.healthline.com/health-news/americans-spend-billions-on-vitamins-and-herbs-that-dont-work-03191.

"The Science Behind Prevagen." Prevegen, http://www.prevagen.com/research/.

"The State of Healthcare: From Challenges to Opportunities." *Issuu*, April 19, 2015, https://issuu.com/sustainia/docs/state_of_healthcare.

14

THE IMPAIRED INDUSTRY OF PRODUCING POWER

> No one gives a damn about how much coal, oil or gas they use—they care about how hot their shower is and how cold their beer.
> —*Amory Lovins*

Amory might have been right. It seems that most people are just interested in flipping the switch, topping off the tank, filling the shopping cart—getting enough power to satisfy their needs, without a lot of afterthought. Let's face it, Americans are energy hogs—the biggest in the world, per capita, and second in its production. The amount of energy wasted in the United States in twelve months could power the United Kingdom for seven years.

With about 5 percent of the planet's population, we gobble approximately more than a quarter of the Earth's energy. China, with more than 1.3 billion people, is the top producer of energy in the world, most of which goes to manufacturing. Coal is China's source of most of that energy, they use almost as much as the rest of the world combined—along with producing the most greenhouse gas (GHG) emissions.

As the world faces the threat of climate change from rising global temperatures, the world's top energy consumers are going to need to reevaluate not only how much energy they use, but also how much they waste. According to the Paris Agreement, signed by 195 countries in 2015 and passed into international law, every country submitted an individual plan to tackle its GHG emissions.

Present POTUS Trump has withdrawn from the agreement claiming that the United States is going to be the cleanest country in the world while also dismantling the EPA, increasing oil drilling on public lands, encouraging coal mining and use, eliminating endangered species in favor of oil drilling and mining, and decreasing the regulations for vehicle emissions, claiming that they are unfair and too expensive. And according to him, "The concept of global warming was created by the Chinese in order to make U.S. manufacturing noncompetitive."

TO UNDERSTAND THE ENERGY SITUATION—A PRIMER

First, laws about energy have to be understood, respected, and followed. The first law of thermodynamics states that energy cannot be created or destroyed. It can only be transferred or converted from one form to another. The second law of thermodynamics tells us that achieving perfect thermal efficiency is as impossible as unscrambling an egg. So, energy waste is universal and we should accept that and realize that there is always going to be a certain amount of energy wasted. But that doesn't mean that waste shouldn't be profligate and can't be mitigated.

Up until the mid-1800s, wood supplied nearly all of the nation's energy needs. The three major fossil fuels—petroleum, natural gas, and coal—have dominated the U.S. energy mix for more than 100 years, starting with the Industrial Revolution when we began using nonrenewable fossil fuels.

Unlike fossil fuels, renewable sources of energy do not directly emit GHGs. In the very near future, as technology increases and prices decrease, the use of alternative sources of energy will amplify. At present, renewable sources are not always available and are difficult to store. For example, clouds shroud the Sun from producing solar power, less wind reduces power from wind farms, and droughts reduce the water available for hydropower. And energy is not easy to store.

Facts and Figures:

- America wastes more than 60 to 80 percent of its energy.
- Most power plants in the United States are only 33 percent energy efficient. This rate has not changed since the 1950s.

- If every American home replaced just one light bulb with an Energy Star bulb, the amount of energy saved could light more than three million homes for a year.
- About 75 percent of the energy used to power our electronic devices is consumed when the products are turned off.
- A quarter of your heating and cooling bill could be going out the window each month—literally. About 25 percent of the air inside your home or business can escape through single-glazed windows.
- The average car emits three times its weight in CO_2 every year.
- Approximately 30 percent of the energy used in most buildings today is either unnecessary or used inefficiently.
- The average household wastes $150 per year in energy costs just from holes and cracks in their homes.

The energy industry is the largest enterprise—$600 trillion annually—in the world. But like any huge endeavor small missteps, carelessness, and inattention can crack and crumble any great enterprise. Each and all of us is accountable as a curator, not only for the present, but for the future building blocks of energy for our country and the world.

A BRIEF HISTORY OF WASTE

In 1970, the U.S. economy actually managed to utilize more energy than it wasted, using 31.1 quadrillion Btus and only wasting 30.6 quadrillion, achieving an energy efficiency higher than 50 percent. (The average U.S. home uses 10,766 kilowatt hours [kWh] per year. One quad is equal to 293 billion kWh or, 183 million barrels of oil). Since then, the overall energy efficiency of the economy has steadily fallen. Power plants and internal combustion engines are notoriously energy inefficient, and as their use has increased, the efficient use of energy has fallen, and transportation has become the number one producer of harmful emissions. In the United States it's estimated that cars, trains, and planes are 21 to 29 percent efficient, on average, and heating, cooling, and lighting are on average 65 percent efficient. Each year, the Lawrence Livermore National Laboratory releases an analysis of the energy input and use of the economy to determine our energy efficiency. Of the 93.6 quadrillion Btus of raw energy that poured into the U.S.

economy in 2016, only 37.0 quadrillion Btus were actually used, the other 58.7 quadrillion Btus were wasted, according to the Energy Information Agency (EIA).

A Primary Energy Primer

The four major nonrenewable energy sources are approximately:

- Petroleum (40 percent)
- Coal (20 percent)
- Nuclear energy (10 percent)
- Natural gas (30 percent)

The common renewable energy sources are approximately:

- Methane from municipal solid waste (21 percent)
- Biofuels, biomass, methane (26 percent)
- Hydropower (24 percent)
- Geothermal (2 percent)
- Wind (21 percent)
- Solar (6 percent)

What's a "Quad"

A "quad" is one quadrillion (a thousand trillion) Btus. Sixty-one percent of the 93.6 quads energy that flows through our economy is ultimately wasted. Energy sources are measured in different physical units: liquid fuels in barrels or gallons, natural gas in cubic feet, coal in short tons, and electricity in kWh. In the United States, Btu, is a measure of heat energy and is commonly used for comparing different types of energy. A Btu is defined as the amount of heat required to raise the temperature of one pound of water by one degree Fahrenheit. In 2016, total U.S. primary energy consumption was about 1,000 trillion Btus.

Here, are a few things equivalent to a quad:

- 8,007,000,000 gallons (U.S.) of gasoline
- 293,071,000,000 kilowatt-hours (kWh)
- 36,000,000 metric (2,204.6 pounds) tons of coal

- 970,434,000,000 cubic feet of natural gas
- 25,200,000 metric tons of oil
- 252,000,000 tons of TNT
- 13.3 tons of uranium-235 (about the size of a large nuclear bomb)

Not So Sunny Waste

The biggest waste of energy, of course, is the Sun, but thankfully, there is enough solar energy available to fulfill all our energy requirements now and forever. It radiates 380 billion-billion megawatts of energy every second, more than humankind has ever used. The Sun is our most usable, inexpensive, and nonpolluting resource. Solar energy can be active, as in photovoltaic cells that produce electricity, or passive, as when sunlight is used to heat water.

The Earth is a huge energy storage device that absorbs 47 percent of the Sun's energy daily or 800 trillion terawatts of power annually. The total world demand for electrical power is 20.3 terawatts per year. Studies have shown that annual energy costs can be reduced by as much as 80 percent by using different solar energy methods, but so far, more than 99 percent of the Sun's energy is wasted.

Wind, like sunshine, can be fickle, and it's tough to determine how much is lost, because it's everywhere, from tiny whiffs to fairly dependable gusts. By 2017, the United States had over 82 gigawatts (a gigawatt is one billion watts) of installed wind power capacity, but was only about seventh in the world for harnessing eolic (wind) power.

Power Wasted Producing Power

Power plants, on average, operate at 33 percent fuel efficiency, meaning that for every unit of electric power generated, two units of waste heat are dispersed. Waste heat is akin to heat produced by a light bulb—it doesn't help illuminate, but wastes energy through heat. In transportation let's assume that we get 30 to 35 percent of the useful energy out of gas (this is higher than a passenger car during normal driving conditions, but probably low for rail, shipping, and long-haul trucking). So, at 33 percent, that implies about 67 percent of waste, going into tailpipe exhaust, brakes, tires, friction, and so on.

Then there is loss of energy due to at least 15 percent inefficiency of the U.S. electrical grid that distributes electrical power to the country. The United States might be the first in consumption per capita, but it was ranked eighth among the world's 23 top energy-consuming countries in efficiency. The prevention of energy waste is one of the most powerful resources we have for meeting our energy and environmental goals. It is also an enormous economic opportunity.

In commercial buildings, where electricity costs are roughly $190 billion a year, about 30 percent of this energy goes to waste. Commercial buildings account for 36 percent of all U.S. electricity consumption and are responsible for 18 percent of carbon dioxide emissions.

The inefficiency of electric power generation (fueled by petroleum sources) also produces a great deal of the pollution in the country. Some of the waste by-products are:

- 63 percent of sulfur dioxide emissions (SO_2), which contributes to acid rain.
- 22 percent of nitrous oxide emissions, which contribute to urban smog, and acid rain.
- 39 percent of CO_2 emissions, which contribute to global climate change.
- 33 percent of mercury emissions, which pose significant health risks to living things.
- Particulate matter (PM) results in hazy conditions in cities and scenic areas, and coupled with ozone, is a respiratory and overall health hazard.

However, power from U.S. electricity is slowly but steadily decarbonizing, and the sector's carbon dioxide emissions fell by 4.9 percent in 2016, while coal use declined by 8 percent.

THE GROANING "GRID" OF WASTE

The U.S. National Power Grid is a Frankenstein-like creation, grafted and sutured together on an outdated electrical framework, assembled via early 20th-century engineering technology. The country's copper-based non-smart electric grid is estimated to leak electricity at around

15 percent and in some instances, it can be as much as 60 percent, especially when power plants are located far from where the electricity is needed. Our electric power system operates at approximately 33-percent overall efficiency, and the rest is lost to waste.

This system has faithfully kept us out of the dark for some time, but it's succumbing to senescence, decay, overload, and waste, and there is literally no light at the end of the tunnel. In 2014, the country's electric grid was losing power more often than any other developed nation, suffering more than 100 national outages per year. Customers in Japan, for example, averaged 4 minutes of downtime every year, while customers in the Pacific Northwest averaged about 214 minutes of downtime. The Department of Energy indicated that these power outages could be costing companies as much as $150 billion per year. And repair or replacement could cost as much as $2 trillion.

Push for More Power

Many factors have forced a slowdown in the construction of new U.S. power plants, especially in the case of coal—a source seen as friend and foe. Fear of nuclear power, especially after the 2011 Japanese tsunami disaster, is uncalled for, but has been revived in many people. In addition, the NIMBYs (not in my backyard) and the BANANAs (build absolutely nothing anywhere near anything) have issues about noise, microwave stations, GHG emissions, transmission lines, increased traffic, transportation of hazardous substances, terrorism, animal deaths, and ruining property values. Against this backdrop are the escalating needs of an ever more energy-hungry public and those who, at the same time, are trying to conserve and reduce the vast amount of power wasted in Americans' homes and offices.

Many new coal-powered plants have been proposed, but the majority have been canceled, shelved, or converted to cleaner and more efficient natural gas. More than 100 orders for nuclear power reactors, many that were already under construction, were canceled in the 1970s and 1980s. Up until 2013, there had been no new groundbreakings for nuclear reactors. Critics call nuclear slow, overbudget, and economically untenable, compared to solar. Since 1956, the cost of nuclear power has gone up by a factor of three, and the cost of solar has dropped by a factor of 2,500.

The Energy Waste Race

The United States is number one—in the category of energy waste. No single person or enterprise can take credit—every industry has aided in filling the energy wastebasket of this country, from the family that leaves all the lights on when out of the house, to the operators of power plants. According to EPA estimates, the transportation and industrial sectors consume and waste more energy than any other sector. And energy use by the residential, transportation, and commercial sectors has drastically increased each year for the past 60 years, and it continues to rise.

U.S. Consumption vs. the World

The poorest 10 percent of the United States accounted for just 0.5 percent of all energy consumption, and the wealthiest 10 percent accounted for 59 percent. If undeveloped countries consumed at the same rate as the United States, four complete planets the size of the Earth would be required for the resources, according to the World Bank economic indicators. Each person in the industrialized world uses as much commercial energy as 10 people in the developing world. The average American wastes about 283 kWh of energy per month, the same as running an electric oven at 350 degrees for six days.

The five percent of the world's population that lives in the United States has more environmental impact than the 51 percent that live in the other five largest countries. From a consumption perspective, the developed countries have a bigger population-growth problem than the undeveloped ones. The next time you hear about a woman in India who has seven children and you complain about the waste these "extra" kids cause, remember that she would have to have more than 20 children to match the impact of the energy and waste of just one American offspring, according to the *Statistical Review of World Energy 2007*.

Population isn't just a matter of the number of people, it's about the energy we need and what we waste to produce everything we need for daily life. In the next 30 years, the population of the United States is projected to increase by nearly 130 million people—the equivalent of adding another four states the size of California, the most populous state in the union.

U.S. Report Card

- The United States gets an F for energy efficiency. This factors in energy wasted in everything from electricity power plants and cars to home appliances.
- Americans spend $130 billion a year on wasted energy that powers standby appliances and heat that warms unoccupied rooms.
- It would cost $1 billion to retrofit 75,000 average-sized homes to save the same amount of energy that would have been generated with the 50 million gallons of oil that leaked into the Gulf of Mexico by British Petroleum, according to a study from Energy Savvy.

THIS IS WHAT WE CAN DO

Federal programs like the Energy Star effort estimates it has saved consumers and businesses $430 billion on their utility bills since its launch in 1992, American families as much as $500 a year, and it has grown to 16,000 partners (including a wide range of manufacturers, retailers, builders, and utilities); a program that Trump wants to shut down in accordance with his controversial stand on climate change.

The result of the inefficiency of electric power generation (fueled by petroleum sources) is second in producing pollution in the country.

- The United States gets the majority of its total energy from oil, coal, and natural gas, according to the Institute for Energy Research .
- 20 percent is wasted in commercial and residential buildings, and 20 percent is wasted in industry or manufacturing.

Here are more places to look for energy savings. We're all guilty of bad habits. Don't live in denial, it only wastes energy. The good news is you can become a reformed energy hog by avoiding power-guzzling habits. Small changes will add up to make a huge difference.

Look for energy-efficiency in household appliances and equipment. The Energy Star ratings are a good way for consumers to improve their energy usage. The federal Energy Star effort estimates it has saved consumers and businesses $430 billion on their utility bills since its launch in 1992 and American families as much as $500 a year.

- A major culprit across all industries is heat waste, the by-product of inefficient technology. Since the sole purpose of a light bulb is to produce light, all of the energy that goes into producing the heat is a complete waste. This is part of the reason why our government is slowly phasing out incandescent bulbs and encouraging the purchase of more efficient bulbs, like compact fluorescent lamps (CFLs). A light fixture with two conventional 60-watt bulbs, left on for eight hours each weekday, would add $25 a year to an electric bill. Perhaps saving $25 might not seem like much, but think of the energy saved if we all—neighborhood, cities, counties, states—started to conserve.
- If you think you might forget to turn the lights and devices off, use a smart home system to remotely monitor your home energy use.
- Look for a programmable timer and thermostat controls. Each one-degree increase of the thermostat setting will save about 10 percent on your annual energy bill.
- Install ceiling fans, they are much cheaper than air conditioning and have less impact environmentally.
- Any home with an HVAC unit has air filters that need to be regularly cleaned; once the air filter clogs, the HVAC expends more energy pulling in air. So to reduce energy use, replace filters every three months.
- Turn on the air conditioner early and try to use it only on really hot or humid days. Begin cooling early rather than waiting until your home is hot. Similarly, start heating early when expecting a cold day.
- When public transportation is not an option, carpool with your neighbors, ride your bike or walk, and combine shopping trips (maybe with friends) to limit drive time and fuel use.
- Shut off computers, printers, and other energy vampires at the end of every day. A cable box with DVR, plugged in for a year and never used, wastes $38 a year.
- Buy appliances with a good energy rating and think about size first. Often it's easier for a larger model to be more efficient (and therefore have more stars) than a smaller one.
- Those few seconds staring into the refrigerator add up. Every year, people spend around 10 hours looking at an open fridge or

freezer, accounting for 7 percent of the appliance's total energy use.

- A chest freezer sized below 16.5 cubic feet costs $53 per year. When your freezer is empty, unplug it to save energy and money.
- Pick the right washing machine. Although they usually cost more to buy, most front-loader washing machines save you money over time and use less energy. Almost 90 percent of a washing machine's energy is spent heating water. Cut energy use in half by switching from hot to warm water, and reduce it even further by using cold water.
- The average dishwasher requires around 1,800 watts of electricity to run at $66 per year, but save about $20 by using it two less days a week and running it only when full. Save around 15 percent of the dishwasher's total energy use by letting the load air dry.
- The Department of Energy recommends 120 degrees for energy efficiency for your water heater.
- Insulate your roof or ceiling. This will help keep your home a pleasant temperature in summer and winter.
- Draft-proof your home by making sure doors and windows are properly sealed—you can buy draft excluders or window seals very cheaply.
- Seal your chimney with a damper. This will help to keep hot or cold air from escaping.
- Close and shade all external windows and doors.
- Water-efficient showerheads are great water-saving devices.
- Check energy use times and try to use less when the demand is greatest.

The average household emits around 14 tons of GHGs every year, half of which is from electricity generation. One simple and relatively cheap way to make a difference is by switching our electricity to "green" power—using energy generated from clean renewable sources. The fact that the United States is the least energy-efficient country in the world may be unbelievable, but it must not be ignored. Each one of us is able to make at least one change in our daily lives to conserve.

SOURCES

"10 Charts That Tell the Story of Energy in 2017." *Forbes*, December 22, 2017, https://www.forbes.com/sites/ucenergy/2017/12/22/10-charts-that-tell-the-story-of-energy-in-2017/#1f38b4f23f75.

"11 Facts about Energy." *Do Something*, https://www.dosomething.org/us/facts/11-facts-about-energy.

"50 Surprising Facts on Energy Consumption in the United States." *Electric Choice*, September 21, 2015, https://www.electricchoice.com/blog/50-surprising-facts-on-energy-consumption/.

Battaglia, Sarah. "US Now Leads in Energy Waste." *The EC Energy Collective*, March 2, 2013, http://www.theenergycollective.com/sbattaglia/193441/us-most-energy-waste.

Borenstein, Seth. "Facts Muddy Donald Trump's Claim U.S. Is Cleanest Country on Earth." *Global News*, June 3, 2017, https://globalnews.ca/news/3500144/donald-trump-climate-change-facts/.

Brown, Claire. "How Cow Poop Could Power the Emissions-Free Hydrogen Car of the Future." *The New Food Economy*, February 8, 2018, https://newfoodeconomy.org/toyota-hydrogen-emissions-free-fuel-cell-car/.

Burgess, James. "How Much Energy Does the U.S. Waste?" *The Cheat Sheet*, September 2, 2013, https://www.cheatsheet.com/money-career/how-much-energy-does-the-u-s-waste.html/?a=viewall.

Burgess, James. "US Wastes Enough Energy to Power UK for 7 Years, Report Finds." *The Christian Science Monitor*, August 28, 2013, https://www.csmonitor.com/Environment/Energy-Voices/2013/0828/US-wastes-enough-energy-to-power-UK-for-7-years-report-finds.

Case, Scott. "The Gulf Oil Spill vs. Home Energy Retrofits." *Energy Savvy*, June 5, 2010, https://blog.energysavvy.com/2010/06/15/the-gulf-oil-spill-vs-home-energy-retrofits/.

Casten, Sean. "How Much Energy Does the U.S. Waste?" *Grist*, September 12, 2009, https://grist.org/article/2009-09-11-how-much-energy-does-the-us-waste/.

"China Produces and Consumes Almost as Much Coal as the Rest of the World Combined." U.S. Energy Information Administration, March 2014, https://www.eia.gov/todayinenergy/detail.php?id=16271.

"Clean Renewable Energy Can Meet All Our Needs." *Energy Justice Network*, http://www.energyjustice.net/solutions/factsheet.

Clift, Jon, and Amanda Green. "10 Facts on United States Energy Use." *Huffington Post*, May 25, 2011, https://www.huffingtonpost.com/2008/08/08/10-facts-on-united-states_n_117074.html.

"Coal and Jobs in the United States." Wikipedia, https://www.sourcewatch.org/index.php/Coal_and_jobs_in_the_United_States.

"Consumption by the United States." https://public.wsu.edu/~mreed/380American%20Consumption.htm.

"How Does America Rank." *Ranking America*, https://rankingamerica.wordpress.com/how-does-the-united-states-rank-in/.

Cusick, Daniel. "Fossil Fuel Use Continues to Rise." *Scientific American*, October 25, 2013, https://www.scientificamerican.com/article/fossil-fuel-use-continues-to-rise/.

Dennis, Keith. "Electricity as the End-Use Option." *Science Direct, The Electricity Journal*, 2015, https://www.sciencedirect.com/science/article/pii/S104061901500202X.

Dondero, Jeff. *The Energy Wise Home*. Lanham, MD: Rowman & Littlefield, 2017.

Dondero, Jeff. *The Energy Wise Workplace*. Lanham, MD: Rowman & Littlefield, 2107.

Dorn, Jonathan. "The End of an Era: Closing the Door on Building New Coal-Fired Power Plants in America." Mary Jane's Farm Girl Connection, 2009, http://www.maryjanesfarm.org/snitz/topic.asp?TOPIC_ID=31455.

"Emissions Trading." Wikipedia, https://en.wikipedia.org/wiki/Emissions_trading.

"Electricity Customers." US Environmental Protection Agency, https://www.epa.gov/energy/electricity-customers.

"Energy Newslinks—Rochester, NY Area." Rochester Environment, http://rochesterenvironment.com/Issues/Energy_NewsLinks.html.

"European Commission." Paris Agreement, 2015, https://ec.europa.eu/clima/policies/international/negotiations/paris_en.

Follet, Andrew. "Lights Out: The Top 7 Threats to America's Power Grid." *Daily Caller*, January 10, 2016, http://dailycaller.com/2016/01/10/top-7-threats-to-americas-power-grid/.

"Green Energy." Malabar Farm, http://www.malabarfarm.org/park-info-and-maps/green-energy.

Gunther, Marc. "Killing Energy Star: A Popular Program Lands on the Trump Hit List." *Yale Environment 360*, May 4, 2017, https://e360.yale.edu/features/killing-energy-star-a-popular-program-lands-on-the-trump-hit-list.

Gupta, Akanksha. "American Wind Energy Association." AWEA Media Center Press Release, March 5, 2104, https://www.awea.org/MediaCenter/pressrelease.aspx?ItemNumber=6184.

"Heat and Cool Efficiently." Energy Star, https://www.energystar.gov/index.cfm?c=heat_cool.pr_hvac.

"How Many BTU Are Required to Raise the Temperature of One Gallon of Water One Degree Fahrenheit In One Hour?" *Answers.com*, http://www.answers.com/Q/How_many_BTU_are_required_to_raise_the_temperature_of_one_gallon_of_water_one_degree_Fahrenheit_in_one_hour.

"How Much Electricity Does an American Home Use?" Energy Information Administration, https://www.eia.gov/tools/faqs/faq.php?id=97&t=3.

"How Much Electricity Is Lost in Transmission and Distribution in the United States?" Energy Information Administration, https://www.eia.gov/tools/faqs/faq.php?id=105&t=3.

Hum, Avery, and Daniel Abery. "Energy-Efficiency-Grade Law Deserves an F." *Crain's*, February 12, 2018, http://www.crainsnewyork.com/article/20180212/OPINION/180219986/letter-to-the-editor-energy-efficiency-grade-law-deserves-an-f.

"What Are the Different Types of Renewable Energy?" Energy Information Administration, https://www.eia.gov/energyexplained/?page=renewable_home#tab2.

Koerth-Baker, Maggie. "Can America Turn Its Nuclear Power Back On?" *Popular Mechanics*, January 21, 2016, https://www.popularmechanics.com/science/energy/a18818/can-us-nuclear-power-get-un-stuck/.

"Laws of Thermodynamics." Wikipedia, 200&, https://en.wikipedia.org/wiki/Laws_of_thermodynamics.

Lehnardt, Karin, "53 Interesting Facts about Energy." *Fact Retriever*, June 30, 2017, https://www.factretriever.com/energy-facts.

Mooney, Chris. "The Supreme Court Could Block Obama's Climate Plans." *Washington Post*, February 10, 2016, https://www.washingtonpost.com/?utm_term=.1244d3874e92.

Mulkern, Anne C. "Trump Seeks More Oil Drilling; in Calif., That's Not So Easy." *E&E News*, December 7, 2016, https://www.eenews.net/stories/1060046782.

"Nuclear Power in the United States." Wikipedia, https://en.wikipedia.org/wiki/Nuclear_power_in_the_United_States.

"Myths about Energy in Schools." Environmental Protection Agency, https://www.nrel.gov/docs/fy02osti/31607.pdf.

Passel, Jeffrey, and D'Vera Cohn. "U.S. Population Projections: 2005–2050." Pew Research Center, February 11, 2008, http://www.pewhispanic.org/2008/02/11/us-population-projections-2005-2050/.

Population and Energy Consumption." *World Population Balance*, 2018, http://www.worldpopulationbalance.org/population_energy.

"Quad Unit." Wikipedia, https://en.wikipedia.org/wiki/Quad_unit.

Roberts, David. "American Energy Use, in One Diagram." *Vox*, April 13, 2017, https://www.vox.com/energy-and-environment/2017/4/13/15268604/american-energy-one-diagram.

Ruhl, Christof. "Energy in Perspective." *BP Statistical Review of World Energy 2007*. June 12, 2007, https://www.bakerinstitute.org/media/files/event/1acc88b2/Energy_In_Perspective-BP_07_slides.pdf.

Savitz, Eric. "America: The Worldwide Leader in Wasting Energy." *Forbes*, February 22, 2013, https://www.forbes.com/sites/ciocentral/2013/02/22/america-the-worldwide-leader-in-wasting-energy/#539214261985.

Simon, Robert M., and David Hayes. "America's Clean Energy Success, by the Numbers." Center for American Progress, June 29, 2017, https://www.americanprogress.org/issues/green/reports/2017/06/29/435281/americas-clean-energy-success-numbers/.

Sobolewski, Terry, and Ralph Cavanagh. "Why Is America Wasting So Much Energy?" *New York Times*, November 7, 2017, https://www.nytimes.com/2017/11/07/opinion/bipartisan-energy-efficiency.html.

"Spooky Statistics about Energy and Water Waste." *Energy Resource Center*, October 29, 2013, https://www.erc-co.org/spooky-statistics-about-energy-and-water-waste/.

"Sources of Greenhouse Gas Emissions." Environmental Protection Agency, https://www.epa.gov/ghgemissions/sources-greenhouse-gas-emissions.

Tabuchui, Horoko. "Coal Mining Jobs Trump Would Bring Back No Longer Exist." *New York Times*, March 29, 2017.

"The Energy Policy of China." Wikipedia, https://en.wikipedia.org/wiki/Energy_policy_of_China.

Trump, Donald. "The concept of global warming was created by and for the Chinese in order to make U.S. manufacturing non-competitive." *Twitter*, November 6, 2012, https://twitter.com/realdonaldtrump/status/265895292191248385.

"Today in Energy." Energy Information Administration, https://www.eia.gov/todayinenergy/detail.php?id=13531.

"Transportation Replaces Power in U.S. as Top Source of CO2 Emissions." *Yale Environment 360*, December 4, 2017, https://e360.yale.edu/digest/transportation-replaces-power-in-u-s-as-top-source-of-co2-emissions.

"U.S. Power Outage Statistics." Diesel Service and Supply, March 2017, http://www.dieselserviceandsupply.com/blog/March-2017/US-Power-Outage-Statistics.aspx.

"US Wastes 61–86% of Its Energy." *Clean Technica*, August 26, 2013, https://cleantechnica.com/2013/08/26/us-wastes-61-86-of-its-energy/Werme.

"Energy Flow in the United States—94.6 quads in 2009." *What's Up with That?*" April 10, 2011, https://wattsupwiththat.com/2011/04/10/energy-flow-in-the-united-states-94-6-quads-in-2009/.

"What Are Some Facts about Energy?" *Reference*, https://www.reference.com/science/energy-d5cd8747b45f071f?aq=facts+about+energy&qo=cdpArticles.

"What Is Renewable Energy?" Energy Information Administration, https://www.eia.gov/energyexplained/?page=renewable_home.

"What Is the United States' Share of World Energy Consumption?" Energy Information Administration, 2018, https://www.eia.gov/tools/faqs/faq.php?id=87&t=1.

"Where the Energy Goes: Gasoline Vehicles." United States Department of Energy, https://fueleconomy.gov/feg/atv.shtml.

Wirfs-Brock, Jordan. "Power Outages on the Rise across the U.S." *Inside Energy*, August 18, 2014, http://insideenergy.org/2014/08/18/power-outages-on-the-rise-across-the-u-s/.

15

THE INDUSTRY OF WASTE

God's garden has become man's junkyard.
—*Anthony T. Hincks*

DOES MODERN TECHNOLOGY OWE ECOLOGY AN APOLOGY?

Equipped with possibly the best technology in the world, the United States isn't even in the top ten countries in recycling and waste management. In fact, at best we come at about 11th.

Municipal solid waste (MSW), commonly known as garbage, consists of everyday items we use and then throw away from homes, schools, hospitals, and businesses. The aim of waste management is to prevent pollution and promote reuse and recycling, to dispose of toxic chemicals, and to conserve energy and resources.

Each day, we fill more than 44,000 garbage trucks, each holding about nine tons of trash, feeding the $75 billion industry of waste in the United States. Currently, approximately 20,000 companies are in the industry. Eight of the largest waste management companies account for nearly half of the industry's yearly revenue. For collection, disposal, and mitigation we pay:

- 60 percent for collection
- 25 percent for disposal
- 15 percent for remediation (cleanup)

The collection process is the biggest part of the waste industry, accounting for approximately 60 percent of the industry's revenue. There are various kinds of collections:

- Automatic collection—almost all communities will provide a means for trash pickup, and perhaps recycling.
- Curbside refers to recycling programs that serve households by collecting garbage and recyclables in bags, bins, or carts.
- Opt-in or subscription service is for communities that require some level of household action to initiate curbside pickup and recycling.
- A private hauler is a company that has been contracted by a city, municipality, or an individual, to provide curbside pickup service for trash and recyclables.
- Public haulers are those that are owned and operated by a municipality.

Single-stream collection of recyclables is the practice of collecting commingled recyclable materials all in one container at the curbside. This varies from "dual-stream" and "multistream" collection, which collects recyclables in two or more receptacles.

Waste treatment and disposal is responsible for 25 percent of the revenue in the industry. It's what is done to the waste after it has been collected. Many countries require a proper treatment method for solid waste to reduce its environmental effects. The most established waste treatment methods include: composting, incineration, landfill, and recycling.

About 15 percent of the annual revenue is remediation. This involves cleaning oil spills, ground contaminates, removal of asbestos and lead paint, restoration of strip-mined areas, and processing and neutralizing hazardous waste.

Most cities and towns charge a flat fee for trash service, so most Americans pay little attention to the amount of waste they are discarding, and that can be a problem. Joshua Reno, an assistant professor of anthropology at Binghamton University who studies trash, points out, "By taking our waste away from us so efficiently, it makes us more inclined to dispose more. Although the country has no shortage of space for its trash, even the best-run landfills can stress the environment."

Landfills still have plenty of room to expand. On average, for every year's worth of trash in the United States, landfills add 2.7 years' worth of capacity. The western region in the country currently has the highest number of landfills with 186,346. Even though available landfill sites have dropped by 80 percent in the last few decades, those still available are megasites with more than enough space for many decades.

MANAGING WASTE

Waste disposal is an enormous economic opportunity. About $200 billion a year is spent on solid waste management, according to Mark Dancy, president of WasteZero, one of the nation's largest waste reduction companies. In the United States, where recycling programs have been operating in full force for years, some experts believe the answer to reducing waste lies in charging for its disposal by weight or other metrics.

WasteZero promotes a bag-based "pay-as-you-throw" program. Used in more than 7,000 cities and counties across the country, the pay program charges residents a set fee for each bag they dispose of at a drop-off location. The program has resulted in an average waste reduction of 44 percent and often doubles recycling rates, according to their statistics. It's human nature that when people get charged for a service, they'll probably conserve and use it less.

According to EPA's Landfill Methane Outreach program, as of July 2013, 621 landfill gas-energy recovery programs were operating in the United States The estimates are that these products will power one million homes. While this is a great way to reuse methane emitted from landfills, only about a third of our greenhouse gas is actually turned into electricity. The rest of the gas is either flared (burned off) or isn't recovered at all. The largest methane emitters are oil and gas, agriculture, and waste management. Clearly there is a lot of room for growth when using landfill gas waste for energy.

Burning

While modern incinerators are not simply toxin-belching behemoths, burning trash often doesn't sit well with many people because of air pollution from chemical toxins and particulates. Before burning, however, scrubbers remove chemicals like dioxins and furans (some of the most toxic chemicals known to science, with no known safe amounts of exposure). And putting less garbage in landfills means less methane in the atmosphere. It also means fewer carbon dioxide emissions from burning fossil fuels. "This gives us the ability to produce electricity from garbage with fewer emissions than from making electricity from coal," says Paul Gilman of Covanta energy, one of the world's biggest companies specializing in waste-to-energy.

But Monica Wilson, program manager at the Global Alliance for Incinerator Alternatives, says these claims are rubbish. "I think they're wrong," she says. "They're turning one problem into a host of others, such as air pollution and a continual reliance on disposable products."

Sustainable Methods

Each year, the EPA produces a report called "Advancing Sustainable Materials Management." It includes information on MSW generation, recycling, and disposal. After 30 years of tracking MSW, the report has been expanded to include additional information on waste prevention, historical landfill info, and information on construction and demolition debris generation.

The report emphasizes the importance of sustainable materials management (SMM), which refers to the use and reuse of materials in the most productive and sustainable ways across their entire life cycle, while minimizing the environmental impacts of the materials we use and making products less toxic. There are many forms of reusable waste products:

- Agricultural and animal wastes include primary crop residues that remain in fields after harvest for fodder, during food, feed, and fiber production.
- Construction and demolition waste includes debris generated during the construction, renovation, and demolition of buildings, roads, and bridges.
- Treatment waste consists of sludge, by-products, co-products, or metal scraps coming from a facility or plant. Sludge is any solid, semisolid, or liquid waste generated from a municipal, commer-

cial, or industrial wastewater treatment plant, water supply treatment plant, or air-pollution control facility.

- Medical and biomedical waste materials are those generated at hospitals, clinics, doctor's offices, dentists, veterinarians, blood banks, home health care facilities, funeral homes, medical research facilities, and laboratories.
- Special wastes are cement dust, mining waste, oil and gas drilling mud, oil production brine, processing waste from phosphate rock mining, uranium waste, and utility or fossil fuel combustion waste.
- Mixed waste includes those that contain radioactive and hazardous waste components making them complicated to dispose and regulate.

Super Cleanup

Approximately 70 percent of Superfund cleanup activities historically have been paid for by parties responsible for the cleanup of contamination. Superfund is a federal government program designed to fund the cleanup of sites contaminated with hazardous substances and pollutants. Until the mid-1990s, most of the funding for cleanup activities led by the government came from a tax on the petroleum and chemical industries. Currently, virtually all funding for government-led cleanup sites under Superfund comes from general revenues or special accounts funded through settlements with potentially responsible parties (PRP). The Superfund program has experienced flat or declining budgets since 2009. Currently, of the $7 billion needed to clean up present sites, only $4 billion has been funded.

Disposal

Before the 1980s, most of the MSW was either buried or burned. During the 1960s and 1970s 90 percent of the municipal solid waste was landfilled with less than 7 percent of materials recovered. The 1980s were more enlightened, and landfill disposal declined to about 54 percent while resource recovery increased to around 33 percent. On average, it costs $30 per ton to recycle trash, $50 to send it to the landfill, and $65 to $75 to incinerate it.

Waste Management Legislation Stats and Facts

The EPA regulates all waste material under the 1976 Resource Conservation and Recovery Act (RCRA). Checkout the following, you may want to use them.

- RCRA of 1970 provides state and local governments with technical and financial help in planning and developing resource recovery and waste disposal systems.
- Used Oil Recycling Act of 1980 and the Used Oil Management Standards of 2003 define the terms used oil, recycled oil, lubricating oil, and re-refined oil.
- Solid Waste Disposal Act Amendments of 1980 target hazardous waste dumping.
- Superfund Amendments and Reauthorization Act (SARA) of 1986 amends the Comprehensive Environmental Response, Compensation, and Liability Act (CERCLA) of 1980 dealing with unacceptable hazardous waste practices, increases state involvement in the Superfund program, and encourages greater citizen participation in decision-making.
- Medical Waste Tracking Act of 1988 defines medical waste and introduces management standards for its segregation, packaging, labeling, and storage.
- Ocean Dumping Ban Act of 1988 prohibits all municipal sewage sludge and industrial waste dumping into the ocean.
- RCRA cleanup reforms of 1999 and 2001 accelerate the cleanup of hazardous waste facilities regulated under RCRA.
- Emergency Planning and Community Right-to-Know Act (EPCRA), also known as SARA Title III of 1980, provides for notification of emergency releases of chemicals, and addresses communities' right to know about toxic and hazardous chemicals.
- RCRA Expanded Public Participation Rule of 1996 encourages communities' involvement in the process of permitting hazardous waste facilities and expands public access to information about such facilities.
- Pollution Prevention Act of 1990 requires the EPA to establish an Office of Pollution Prevention and the owners and operators of manufacturing facilities to report annually on source reduction and recycling activities.

- Hazardous Waste Combustors; Revised Standards; Final Rule—Part 1 of 1998 provides for a conditional exclusion from RCRA for fuels that are produced from a hazardous waste and promotes the installation of cost-effective pollution-prevention technologies.

Shipping Away Waste

We send rawhides to China and get back shoes, waste paper and get back packaging, scrap metal and get back machinery, raw cotton and get back finished clothes. According to Drewry Shipping Consultants, out of 100 containers shipped from China, 60 go back to China empty. That's because one our greatest importers of garbage, the Chinese, are cutting back, causing concern to our export garbage industry. They claim that at least 24 categories of our recyclables and solid waste products are a threat to their environment and public health. Other countries are starting to levy taxes for importing garbage.

There might be a refuse-lined cloud from super-recyclers like Sweden. Less than 1 percent of their household waste ends up in a rubbish dump, so importing garbage as raw material might be good business for them and for us. However, their need would be far less than from a country like China.

There is almost nothing as hard to recycle as electronics. While American firms are supposedly trying to dispose of old devices like phones and computers in an environmentally safe way, other businesses cut costs by shipping them to other nations. (See chapter 12 on E-Waste.)

TRASH INTO CASH

Landfills are now being considered more for their long-term assets than their negative reputations. The Organization for Economic Co-operation and Development (OECD) says 3 billion tons of trash a year will contribute to landfills worldwide by 2030, up from 1.6 billion in 2005. However, the potential for this to be a vast new resource, as opposed to a useless burden, is encouraging.

Landfills are not only a largely untapped resource for many precious metals, but also for recyclable materials formerly regarded as waste that can be mined from landfills. In the United States, more than 4.6 million tons of electronic waste were disposed of in American landfills in 2000 and are sitting, waiting to be located and reused. Such material has the potential to provide a new source for declining and/or expensive supplies of metals such as platinum, vanadium, gold, silver, and copper.

The concept of "by-product synergy" (BPS) consists of taking the waste stream from one production process and using it to make a new product. So, while waste is increasing, so is the cost of raw materials. Productively using waste instead of trashing it can cut costs by reducing disposal fees, reducing the use of resources and greenhouse gases, and opening up additional revenue streams through by-product sales.

Cast-off agricultural waste (e.g., corn husks or scraps from dining halls) can help close the loop in animal feed cost. For example, cattle takes 12 pounds of grain, 2,500 gallons of water, and 35 pounds of topsoil to produce one pound of beef. Another example is a cement manufacturer using slag from a neighboring steel mill in its production process, increasing production output by 10 percent and decreasing nitrogen oxide emissions by nearly 40 percent.

Passing Gas

Like the human stomach, anaerobic (involving, or requiring an absence of free oxygen) digestion facilities use microbes to break down organics into biogas, primarily from methane and carbon dioxide. Like our stomachs and intestines, machines grind organics into a slurry the consistency of a milkshake, which is fed into large, airtight tanks and heated to about 100 degrees, called a digester. The result is biogas used as fuel for vehicles or converted in a power plant to create electricity. Estimates are that anaerobic digestion will cost around $35 to $50 per ton, comparable to less than the average price of burying a ton of city waste in landfills.

The number of landfill-to-energy projects has grown from 21 in 2000 to 621 in 2013, according to the EPA, and as of 2018, there are 632 operational landfill gas (LFG) energy projects in the United States and approximately 510 landfills that are good candidates for projects, an increase that the EPA says would create enough power to serve 700,000 homes. LFG is composed of roughly 50 percent methane (the primary component of natural gas), 50 percent carbon dioxide, and a small amount of nonmethane organic compounds.

Revenue from selling the fuel doesn't come anywhere close to covering its costs yet. But producers benefit from a generous subsidies package. When all the subsidies are tallied up, it's about $1 to $1.50 per gallon cheaper than gasoline or diesel.

Profit in Corporate Waste

According to Rick Perez, chairman and CEO of Houston-based Avangard Innovative, there's profit to be found in factory waste bins. "You've already paid for the packaging on items you've bought," says Perez, "so you'll get 100 percent profit if you turn this waste into something useful. We turn waste into money, something most companies don't focus on."

When he saw that discarded "one-way" plastic bottles would soon start flooding landfills he saw green. A trend at the time was an increase in the demand for synthetic fibers used in the manufacture of polyester-based products. His company was soon processing 94,000 tons of plastic bottles and became the largest recycler of plastic bottles in the world.

Next was supermarket chains, grocery stores, and other retailers, all with similar dilemmas around their waste packaging. He claims that a typical grocery store chain can achieve 20 percent net profit through waste recycling and improved efficiencies. "Anything we can do that takes away from dumping things into landfills and acknowledges the importance of the environment, is a good goal." he says. Most companies view sustainability programs as a cost, but Perez sees exactly the opposite—an opportunity to make money.

Bionic Yarn is a company that has transformed millions of bottles recovered from shorelines, turning them into yarns and fabrics for clothing. They also make fabric for performance or industrial applications—like snowboarding jackets and window shades.

One obstacle is that many older landfill sites may be inaccessible for mining operations. Another challenge is the lack of federal landfill-mining laws and uncertainty about how the process might be regulated. And there is always a risk when bidding on something sight unseen. Companies want assurance that they will turn a profit, and that depends on what is buried unseen in a landfill.

Representatives of Ecomaine, a southern Maine landfill, estimate that since late 2011, nearly 27,000 tons of metal worth more than $2.3 million have been mined from its landfill, which exists on 240 acres over parts of Westbrook, Scarborough, and South Portland.

The waste disposal company, which is owned and operated by 21 southern Maine municipalities, also generates nearly 100 megawatts of electricity each year. By the time the landfill facility is mined completely, as much as 45,000 tons of metal are expected to be pulled from the debris, worth approximately $3.9 million.

"By mining landfills, you recover resources that you don't have to go out and mine from the natural world," said an Ecomaine spokesperson, "If you come up with a ton of nickel, that's a ton of nickel you don't have to pull out of virgin land somewhere."

Slug It Out

One day, landfill sites might be mined for valuable metals using genetically engineered slugs or repurposed microorganisms. Dr. John Collins, commercial director of research center SynbiCITE, believes revolutionary gene cell technology called Crispr-Cas9 could herald the ability to create organisms that digest waste and convert it into useful products, or produce a "broth of cells" designed to change color on contact with certain metals. "Biomining is going to be a thing of the near future," says Collins.

Large quantities of lithium that could be mined and reused to create batteries for electric cars lie buried deep within old landfill sites and could be reclaimed potentially with the help of genetically enhanced organisms. "In terms of creating a bacteria that biomines, there are already people doing that on a very small scale in gold mines."

Collins's colleague Professor Richard Kitney says developments in synthetic biology could allow new types of "plastic-eating biological devices to be created . . . non-biodegradable plastics . . . could be turned into biodegradable material."

A Body of Precious Metals

Scientists are now looking into the possibilities of "poop mining." Researchers at the U.S. Geological Survey (USGS), along with scientists at

Arizona State University have measured gold, silver, platinum, copper, zinc, and other precious industrial metals in biosolids.

Gold could be coming from food products (actually biosolids), dental fixtures, and from medical facilities. Gold and silver is used to treat arthritis and cancer as well as in some surgical and diagnostic procedures, little flakes of gold and silver from jewelry can enter wastewater when a person does the dishes or takes a shower, and precious particles that are used in a variety of consumer products due to their antibacterial properties go down the drain and are flushed out at wastewater treatment plants.

Concentrations of some metals in the biosolids material—say about one part per million of gold, for example—can exceed naturally occurring metal concentrations in earth. There's money in them thar sewage-treatment pools and piles of rock. Near mining sites, piles of waste are left behind. This waste rock and drainage waters could contain metals with concentrations that were too low to be economically recoverable at the time, or metals that weren't of interest then, but that now have new high-tech applications.

Road to Riches

Every day on the road, cars eject particles of platinum, palladium, and rhodium from their catalytic converters. More than $98 million worth of precious metals accumulate on British roads every year, making their roads a low-grade mining opportunity—and you don't have to go underground, as the stuff is sitting on the surface, just waiting to be collected. The UK has about 240,000 miles of paved road, the United States has 4.12 million miles. Do the math.

Technology That Clears the Air

Pyrolysisis considered a waste-to-energy technology; a thermochemical anaerobic decomposition of organic and some inorganic material at elevated temperatures, which means it doesn't release harmful contaminants into the atmosphere that are difficult if not impossible to capture.

The pyrolysis technology, with ocean-waste capture operations along with the mining and processing of waste at landfills, can result in a massive reduction and of pollutants, especially plastics, and the transforming of garbage into a precious resource. By-products include synthesis gas used in:

- Heat production for drying or industrial purposes.
- Industrial steam or thermal oil production.
- Replacement of conventional fuel in the existing boilers.
- Electricity generation, which converts thermal energy into electricity, a process similar to a steam turbine, but in lower temperatures and uses refrigerant instead of water.
- Biochar, used in agriculture to enhance soil fertility, improve plant growth, and provide crop nutrition—improving overall farming productivity. It has also gained considerable attention as animal feed.

Pyrolysis oil is used for:

- Fuel for further refining.
- Food aromas like liquid smoke.
- Pesticides and plant enhancers.
- Torrefied fuel (which converts biomass into a coal-like material that has better fuel characteristics than the original biomass).
- Co-firing in existing coal-fired power stations.
- Alternatives and replacement for conventional fuels.

WHAT CAN WE DO WITH CO_2?

What has been one of mankind's latest and potentially most perilous nemeses might prove to be one of mankind's most modern and unforeseen friends. One of humanity's dreams is to mimic plant life by turning the most ubiquitous waste product on Earth—carbon dioxide—into useful products. We might be on the cusp of a critical method of productively using a hazardous greenhouse gas.

Carbon dioxide is left over after burning biomass and petroleum products for energy. Carbon dioxide combined with water vapor, including smaller amounts of greenhouse gases, methane, and nitrous oxide, acts as a thermal blanket for the Earth, absorbing heat and warming the surface for life support. It acts as a kind of a thermostat, but when the thermostat changes radically it can bring on an ice age or a hot house age that can have deadly effects on the planet's temperature, and living things.

Discounting emissions from agriculture and the built environment, oil, gas, and coal companies account for 71 percent of man-made greenhouse waste gases put into the atmosphere. Just 25 companies are responsible for more than half of industrial emissions since 1988, according to the Climate Accountability Institute. Fossil fuel companies released more emissions in the last 30 years than in the previous 250. If emissions remain at the same rate, by 2050 it might be adios America, and everyone else, for that matter, as those greenhouse gases push us toward adapting to radical climate change or extinction. And the rate at which we're causing change in the environment, we might not have the time to adapt.

Governments have spent billions funding carbon capture, and sequestering (CO_2 storage), but with little luck. Basically, major CO_2 producers like Duke Energy (one of the largest electric power companies in the United States) promised sequestering CO_2 for its plants years ago—but in reality they claim the technology is just too expensive. Also some scientists claim that the sequestered gas can seep back onto the surface.

However, there's a new sheriff in town—a range of potential products that can be produced from CO_2 that could mitigate the amount of gas going into the atmosphere. A recent competition organized by the Canadian-based Change and Emissions Management Corporation, and funded by the region's heaviest emitters, showcased 24 good ideas in its $35 million global prize challenge for "innovative carbon uses."

When it comes to dealing with CO_2 emissions, one of the more ambitious, but still largely unproven fixes, could be a process that sucks CO_2 out of the atmosphere and/or from industrial plants. An international team of scientists working in Iceland has successfully demonstrated that CO_2 emissions can be pumped underground and altered chemically to form a white, chalky, solid calcium carbonate, like in coral, that could be used in making steel, glass, concrete, and paper, among other building materials.

One Canadian startup is turning carbon emissions into pellets that could be used as a synthetic fuel source, while another Swiss company called Climeworks is pumping extracted carbon to farms for agricultural use.

Bulk chemicals already produced routinely from CO_2 include urea to make nitrogen fertilizers, salicylic acid as a pharmaceutical ingredient, and polycarbonate-based plastics. An interesting project is one that involves a process for reducing CO_2 to formic acid (HCO_2H), which has the potential to power fuel cells for electricity generation, other fuels, and chemicals, including some plastic materials.

But like all things that initially sound great, there are a few catches to sucking CO_2 out of the atmosphere and making it manageable. One is that CO_2 is very stable, which means it takes extra effort to activate the molecules so they will react and change. Several researchers are working on "artificial photosynthesis," which involves designing photocatalyst systems, used to speed up reactions in combination with light and solar energy to reduce CO_2 to hydrocarbons. This approach has the potential not only to help alleviate global warming but to augment decreasing petro-fuels.

A new method for taking carbon dioxide directly from the air and converting it to nanoscale (tiny) fibers made of carbon could lead to an inexpensive way to make a valuable building material—and may even serve as a weapon to combat climate change. Carbon fibers are increasingly being used as a structural material in the aerospace, automotive, boat manufacturing, and other industries that value its strength and light weight. It's possible that carbon fiber composites will eventually substitute for steel, aluminum, and even concrete as a building material.

MIT engineer Angela Belcher is now taking a new approach that would remove carbon dioxide from the environment by genetically engineering ordinary baker's yeast; the process converts carbon dioxide into carbonates that could be used as building materials. It has been tested in the lab and can produce about two pounds of carbonate for every pound of carbon dioxide captured.

Thanks to a new technique, ethanol could soon be produced directly from carbon dioxide in the atmosphere. The new method, presented in the Proceedings of the National Academy of Sciences, uses water, carbon dioxide, and electricity delivered through a copper catalyst.

A real heart stopper is a concept being researched at Washington University. They calculate that given an area of 355,200 square miles, their technology could remove enough carbon dioxide to make global atmospheric levels return to preindustrial levels within 10 years, even if we keep emitting the greenhouse gas at a high rate during that period. But don't hold your breath just yet. It would take 10 percent of that area covered with solar panels, using that energy for 10 full years.

The approximately 115 million metric tons of CO_2 used annually by the global chemical industry really doesn't make a ding in the approximately 24 billion metric tons of our annual CO_2 emissions, the huge amount that we're dealing with, especially in a world that hasn't stopped burning coal.

It's great that there are people who are optimistic about finding new ways to utilize it as a raw material. But these products would have to be in heavy demand; otherwise large piles of unwanted "pretty white mountains" of CO_2 calcium carbonate would dot the globe.

All in all, there's no silver-bullet solution to what to do with waste. It's about finding many new and more efficient ways of disposing of our waste cleanly, efficiently, and productively. And it's about changing behavior so that people and societies are encouraged to reduce and recycle waste.

SOURCES

"20 Facts about Waste and Recycling." CB Environment Limited, https://www.cbenvironmental.co.uk/docs/Recycling%20Activity%20Pack%20v2%20.pdf.

"Advancing Sustainable Materials Management: Facts and Figures Report." U.S. Environmental Protection Agency, https://www.epa.gov/facts-and-figures-about-materials-waste-and-recycling/advancing-sustainable-materials-management.

"Artificial Photosynthesis." Revolvy, https://www.revolvy.com/topic/Artificial%20photosynthesis.

"Basic Information about Landfill Gas." U.S. Environmental Protection Agency, https://www.epa.gov/lmop/basic-information-about-landfill-gas.

Barker, Ned. "U.S. Trade with China, Expectations vs. Reality." *Frontline*, 2004, https://www.pbs.org/wgbh/pages/frontline/shows/walmart/china/trade.html.

Cherlin, Edward. "Can We Reverse Global Warming?" *Quora*, April 11, 2016, https://www.quora.com/Can-we-reverse-global-warming.

Dockrill, Peter. "Scientists Have Figured Out How to Turn CO_2 into Solid Rock within Months." *Science Alert*, June 10, 2016, https://www.sciencealert.com/scientists-have-figured-out-how-to-turn-co2-into-solid-rock-within-monthst.

"Food Choices and the Planet." *Earth Save*, http://www.earthsave.org/environment.htm.

"Funded Projects, to Reduce GHG Emissions, and We're Helping Innovators Address Barriers to Commercialization." Emissions Reductions Alberta, http://eralberta.ca/projects/.

Gallego, Jelor. "New Carbon Capture Method Turns CO_2 into Solid Rocks." *Futurism*, June 13, 2016, https://futurism.com/carbon-capture-method-turns-co2-into-solid-rocks/.

Green, Alex, and Sean Bell. "Pyrolysis in Waste to Energy Conversion." Technology Laboratory College of Engineering, University of Florida, 2006, http://www.seas.columbia.edu/earth/wtert/sofos/nawtec/nawtec14/nawtec14-3196.pdf.

"Government in Cahoots with Incinerator Industry." Indymediaireland, November 20, 2006, http://www.indymedia.ie/article/79790.

Grigorvants, Olga. "Turning Garbage into Profit." *Pacific Standard,* September 2, 2015, https://psmag.com/environment/turning-garbage-into-profit.

Higgins, Mark. "Development of Appropriate Sustainable Decisions Support for Disruptive Innovations in Solid Waste Management." Slide Share, https://www.slideshare.net/RJRandall/rgreen-landfill-development-of-tools.

Hornstein, Frank. "Garbage Burning Emits Significant Amounts of Toxic Pollutants, Particulates." *MinnPost*, June 13, 2013, https://www.minnpost.com/community-voices/2013/06/garbage-burning-emits-significant-amounts-toxic-pollutants-particulates.

"How Does the U.S. Manage Their Solid Waste?" *Quora*, https://www.quora.com/How-do-the-U-S-manage-their-solid-waste.

"Industrial Efficiency Challenge." Emissions Reductions Alberta. http://eralberta.ca/.

Ingram, Anthony. "Why Natural Gas Fuel Is a Better Fit for Trucks Than Cars." *Christian Science Monitor*, April 25, 2013, https://www.csmonitor.com/Business/In-Gear/2013/0425/Why-natural-gas-fuel-is-a-better-fit-for-trucks-than-cars.

"IPCC: 30 Years to Climate Calamity If We Carry on Blowing the Carbon Budget." *Guardian*, https://www.theguardian.com/environment/2013/sep/27/ipcc-world-dangerous-climate-change.

"Is Sustainable Trash-Burning a Load of Rubbish?" *Smithsonian*, August 1, 2016, https://www.smithsonianmag.com/science-nature/burning-trash-solution-our-garbage-woes-or-are-advocates-just-blowing-smoke-180959924/.

Jaffe, Eric. "Your Street May Literally Be Paved with Gold, and Other Precious Metals." *City Lab*, June 4, 2013, https://www.citylab.com/life/2013/06/your-street-may-literally-be-paved-gold-and-other-precious-metals/5778/.

Kirk, Mark. "A Billion-Dollar Company Wants to Turn Kindergartners into Entrepreneurs." *Pacific Standard*, December 25, 2017, https://psmag.com/education/turning-kids-into-entrepreneurs.

Koenig, Seth. "Mining Metals from Maine Landfills? Burgeoning Effort Is Already Worth Millions." *BDN Business*, March 14, 2014, http://bangordailynews.com/2014/03/14/business/mining-metals-from-maine-landfills-burgeoning-effort-is-already-worth-millions/.

"Landfill Gas Renewable Energy Fact Sheet." *National Waste and Recycling Association,* 2013, http://www.beginwiththebin.org/images/documents/landfill/Landfill-Gas-Renewable-Energy-Fact-Sheet.pdf.

"Less Than 1% of Sweden's Trash Ends Up in Landfills." IFLScience, http://www.iflscience.com/environment/less-1-swedens-trash-ends-landfills/.

Loeb, Josh. "Genetically Engineered Slugs to Chew Through Landfill and Mine Precious Metals." *E&T*, November 8, 2017, https://eandt.theiet.org/content/articles/2017/11/genetically-engineered-slugs-to-chew-through-landfill-and-mine-precious-metals/.

Maleon, Robert. "World's Worst Waste." *Forbes*, May 24, 2006, https://www.forbes.com/2006/05/23/waste-worlds-worst-cx_rm_0524waste.html#5a64b0663d79.

Mangan, Andrew. By-Product Synergy Networks Offer Zero-Waste Solutions for Businesses, Cities Worldwide." *Sustainable Brands*, February 14, 2013, http://www.sustainablebrands.com/news_and_views/waste_not/BPS-networks-offer-zero-waste-solutions-businesses-cities.

"Metals in Biosolids, Other Soil Amendments, & Fertilizers." *Nebra,* August 28, 2015, https://static1.squarespace.com/static/54806478e4b0dc44e1698e88/t/55e0acdae4b0d463ea047c9f/1440787674598/MetalsInBiosolidsFertilizersSoils-28Aug2015.pdf.

"Mining for Metals in Society's Waste." *The Conversation*, October 1, 2015, https://theconversation.com/mining-for-metals-in-societys-waste-43766.

"Mining Landfills, Strategic Metals: Will Future Supply Be Able to Meet Future Demand?" *Mission 2016, The Future of Strategic Landfills*, 2106, http://web.mit.edu/12.000/www/m2016/finalwebsite/solutions/landfill.html.

Moazed, Alex. “How to Claim a Piece of the $100 Billion Waste Management Opportunity.” *Inc.*, November 28, 2015, https://www.inc.com/alex-moazed/how-to-claim-a-piece-of-the-100-billion-waste-management-opportunity.html.

“Municipal Solid Waste.” Environmental Protection Agency, 2013, https://archive.epa.gov/epawaste/nonhaz/municipal/web/html/.

“New Method to Make Ethanol Out of Carbon Dioxide,” *IFL Science*, http://www.iflscience.com/technology/new-method-to-make-ethanol-out-of-carbon-dioxide/.

“New Report Shows Just 100 Companies Are Source of Over 70% of Emissions.” CDP, July 10, 2017, https://www.cdp.net/en/articles/media/new-report-shows-just-100-companies-are-source-of-over-70-of-emissions.

Orcutt, Mike. “Researcher Demonstrates How to Suck Carbon from the Air, Make Stuff from It.” *MIT Technology Review*, August 19, 2015, https://www.technologyreview.com/s/540706/researcher-demonstrates-how-to-suck-carbon-from-the-air-make-stuff-from-it/.

“Panning for Silver in Laundry Wastewater.” *American Chemical Society/Science Daily*, December 20, 2017, https://www.sciencedaily.com/releases/2017/12/171220121706.htm.

“Pyrolysis.” Wikipedia, 2018, https://en.wikipedia.org/wiki/Pyrolysis.

Puckett, Jim.”The Basel Ban: A Triumph over Business-as-Usual.” Basel Action Network, October 1997, http://archive.ban.org/about_basel_ban/jims_article.html.edu/12.000/www/m2016/finalwebsite/solutions/landfill.html.

Rithvik11. “Burning Fuels Releases Carbon Dioxide, a Green House Gas, Which Causes Climate Change.” *Brainly*, https://brainly.in/question/1845302.

Robert, Jon. “Garbage: The Black Sheep of the Family.” Oklahoma Department of Environmental Quality, http://www.deq.state.ok.us/lpdnew/wastehistory/wastehistory.htm.

Ritter, Steven. “What Can We Do with CO_2?” *Chemical & Engineering News*, April 2007, https://www.researchgate.net/publication/272129947_What_can_we_do_with_CO2.

Rueb, Emily. “How New York Is Turning Food Waste into Compost and Gas.” *New York Times*, June 2, 2017, https://www.nytimes.com/2017/06/02/nyregion/compost-organic-recycling-new-york-city.html.

Shipman, Matt. “From Trash to Cash: Old Landfills Yield New Opportunities.” *Quest*, July 9, 2013, https://ww2.kqed.org/quest/2013/07/09/from-trash-to-cash-old-landfills-yield-new-opportunities/.

Simmons, Ann M. “ The World’s Trash Crisis, and Why Many Americans Are Oblivious.” *Los Angeles Times*, April 22, 2016, http://www.latimes.com/world/global-development/la-fg-global-trash-20160422-20160421-snap-htmlstory.html.

“Solid Waste & Landfill Facts.” University of Southern Indiana, https://www.usi.edu/recycle/solid-waste-landfill-facts.

“Study: Twice As Much Trash Put in Landfills as Estimated.” *Seattle Times*, September 21, 2015, https://www.seattletimes.com/seattle-news/environment/study-twice-as-much-trash-put-in-landfills-as-estimated/.

“The Global Company That Turns Trash into Cash.” *Real Leaders*, https://real-leaders.com/the-global-company-that-turns-trash-into-cash/.

“The Unknown $75 Billion Industry.” *Gridwaste*, August 20, 2014, https://www.gridwaste.com/news/2014/8/20/10l92d52vzaw1zdsf0znubdqpttq7k.

Trafton, Anne. “Scientists Convert Carbon-Dioxide Emissions to Useful Building Materials, Using Genetically Altered Yeast.” *MIT News*, September 22, 2010, https://phys.org/news/2010-09-scientists-carbon-dioxide-emissions-materials-genetically.html.

“Transforming Manufacturing Waste into Profit.” *Working Knowledge*, October 3, 2011, https://hbswk.hbs.edu/item/transforming-manufacturing-waste-into-profit.

“What Is Waste Management?” *Conserve Energy Future*, https://www.conserve-energy-future.com/waste-management-and-waste-disposal-methods.php.

“Where Does E-Waste End Up?” *Greenpeace*, February 24, 2009,.https://www.greenpeace.org/archive-international/en/campaigns/detox/electronics/the-e-waste-problem/where-does-e-waste-end-up/.

Yan, Kimberly. “This Company Turns Plastic Bottle Trash from the Ocean into Clothing.” *Huffington Post*, April 3, 2017, https://www.huffingtonpost.com/entry/this-company-turns-used-plastic-bottles-from-shorelines-into-clothing_us_57d17909e4b06a74c9f301f3.

Yuman, Dan. "Spooked by Scana Failure, Duke Energy Calls It Quits." *The Energy Collective*, August 28, 2017, http://www.theenergycollective.com/dan-yurman/2411699/spooked-scana-failure-duke-energy-calls-guits-le.

Zeller, Tom Jr. "Carbon Capture and Storage: Global Warming Panacea, or Fossil Fuel Pipe Dream?" *Huffington Post*, December 6, 2017, https://www.huffingtonpost.com/2013/08/19/carbon-capture-and-storage_n_3745522.html.

16

GOVERNMENT'S SENSELESS SUPER SPENDING

It is a popular delusion that the government wastes vast amounts of money through inefficiency and sloth. Enormous effort and elaborate planning are required to waste this much money.
—*P. J. O'Rourke, writer and humorist*

Everyone has heard notorious stories of $660 hammers for the Navy, and $640 toilet seats for the Air Force, a $325 million bridge to nowhere, the $65 million plane that never flew, and $12 billion in shrink-wrapped hundreds of cold hard cash on pallets that just "disappeared" in Iraq. Some of them are rumors, exaggerations, or bad bookkeeping, but some of those stories, and a multitude of others are true.

People wonder why there is so much of what seems to be insane fiscal abuse in government. One reason is simply that bumbling bureaucrats, scammers, lobbyists, and politicians seeking votes and favor have lots of experience squandering our tax dollars. And they don't have to worry much because it's not their money, and the federal government has put a fair amount of effort into not knowing what they don't want to know.

Arizona Senator Jeff Flake voted to continue a moratorium on earmarks, what some call "pork-barrel spending," commented, "You can't drain the swamp by feeding the alligators pork." However, members of Congress will always find ways to slather the lard around.

The earmarks for special favors are many times tit-for-tat arrangements between members of Congress to support certain pieces of legislation—you vote for my bill, and I'll vote for yours, and we'll add a little pork for the folks back home. In worst cases, appropriated funds end up lining the pockets of corrupt politicians.

Once in a while, someone is caught with both hands stuck in the cookie jar, and according to the Justice Department's Criminal Division, "When elected officials betray the trust and confidence placed in them by the public, the department will do everything we can to ensure that they are held accountable when caught lining their own pockets." Regrettably, most politicians are too foxy to be caught. And the waste of taxpayer dough goes on.

THE "PORK BARREL" PAST

A political earmark, allocating funds for localized projects secured primarily to bring money to a representative's district, has been a part of American government since the founding of the nation. Thomas Jefferson, for one, expressed grave concern over this clause and wrote to James Madison in 1796 that this was "a source of boundless patronage . . . to members of Congress and their friends, . . . [causing] a scene of eternal scramble among the members who can get the most money wasted in their state." In 1822, President James Monroe said that such expenditures should be limited "to great national works only, since if it were unlimited it would be liable to abuse and might be productive of evil."

"Honest Abe" Lincoln traded Civil War contracts to northern businessmen in exchange for patronage jobs and campaign support. New York in the early 19th century was dominated by Tammany Hall, which controlled politics by favors and purse strings.

Use of the phrase "pork barrel" dates back to 1863, initially in reference to any money that a government or politician would spend on their citizens in exchange for their support. Another possible origin for the phrase comes from practices in the antebellum South in which whites would watch their slaves fight each other over a barrel of salt pork.

More recently, Senator John McCain (R-Ariz.) called it "a corrupt system." And the infamous lobbyist Jack Abramoff, called it the "favor factory." In 1991, Citizens Against Government Waste (CAGW) issued the first Congressional Pig Book, which questioned 546 projects costing

$4 billion. The book contained only a partial listing of what Congress wasted that year on spending that served no real national interest.

Most advocates of these expenditures cite Article 1, Section 8 of the U.S. Constitution, which grants Congress the power to determine how funds collected from taxes should be spent, referred to as "the power of the purse." Perhaps it should be renamed, the power of the pork.

The American public turned against the practice of pork barrel politics toward the end of 2005 in reaction to the infamous $325 million "bridge to nowhere" in Alaska. After public outcry—even from Alaskans—over the blatant exhibition of pork, the funds were rerouted, and the bridge was not built. But a year later, the pork was still rampant and the earmark process was marked by the conviction and jailing of Republican Representatives Duke Cunningham of California and Bob Ney of Ohio, along with congressional staff and others.

Tom Coburn, ex-congressman and senator, has categorized earmarks as "the gateway drug to spending addiction" and said that "restoring earmarks in today's Congress would be like "opening a bar tab for a bunch of recovering alcoholics."

DOUBLE-FISTED FISCAL WASTE

Some spending is controversial in terms of whether or not it is waste. For example, some people believe that funds for public broadcasting is an ill-use of the public's money, others point to statistics that appear otherwise. Fraud, on the other hand, is wrongful or criminal deception intended to result in financial or personal gain—always bad news. The examples noted below are rather outrageous, as well as enlightening, an almost entertaining mixture of foolishness and frivolity—at our expense.

Imaginative excess:

- Every year the Department of Transportation gives Oklahoma $150,000 for an airport that receives one flight a month.
- Vice President Joe Biden and his staff popped in Paris for one night and the hotel bill came to $585,000.
- The U.S. government spent $27 million to teach Moroccans how to design and make pottery in 2012—one of their historically major industries.

- The U.S. Department of Agriculture (USDA) has spent $300,000 to encourage Americans to eat caviar produced in Idaho.
- The federal government spends more than $1.7 billion a year to maintain 770,000 empty buildings, while other agencies are leasing or buying new space.
- A privately owned International House of Pancakes received $765,828 to help build a restaurant in an apparently "underserved" area of Washington, D.C.
- California received $180,000 from a HUD program that's supposed to help spur economic development in lower-income locales—in a ghetto called Beverly Hills.
- The government spent $43 million to build a natural gas station in Afghanistan. Oops—it seems vehicle conversion costs as much as $800—tough in a country where the average annual income is about $690.
- The Department of Commerce and the USDA teamed up to provide over $1 million to help a county in New York build a new yogurt factory for PepsiCo, Inc., hardly an impoverished multinational company.
- $100,000 was allotted for the Coast Guard to patrol some of the country's most exclusive real estate to stop uninvited guests from crashing private parties.
- The 2007 "Big Dig" in Boston, in which a 3.5-mile section of highway was relocated underground, cost nearly $15 billion, or almost 4.3 billion per mile—at $12 billion over budget.
- $41 billion was spent on a missile defense system with a 70-percent failure rate.
- NASA spent $45,000 for experiments on the International Space Station that "will examine a variety of coatings and metals used in golf products." A precursor to Alan Sheppard smacking a couple of balls on the moon in 1971?
- About $350,000 a year is spent on hair care services for members of the U.S. Senate.
- We spent $2,600,000 on "parliamentary strengthening" in Eastern Europe. That means we taught them how to balance and follow a budget—that's like the blind giving driving instructions.

Expenditures to study animal and medical mysteries:

- $387,000 to study the effects robot-provided Swedish massage has on the physical recovery of rabbits after exercise.
- $856,000 to train mountain lions to run on a treadmill in order to measure the energy consumption of the cats' hunting techniques.
- $171,000 to teach monkeys to gamble in order to determine if monkeys, like humans, believe in the concept of a "hot hand."
- $331,000 to study "hunger" by testing whether hungry spouses were more likely to stab a voodoo doll representing their significant other.
- $820,000 to determine the impact of public breastfeeding on the rate of car accidents at crowded intersections.
- $371,000 to study if mothers loved their dogs as much as their own kids by studying the way their brains responded to pictures of both.
- $484,000 to study whether "drunk recall" of information learned while intoxicated is a real phenomenon, as part of a program titled "E = MC hammered."
- $740,000 to Montana State University to research the use of sheep grazing as a means of weed control.
- $50,000 from the USDA to help Alpaca farmers market Alpaca manure, a.k.a. the "perfect poop."
- $505,000 to promote specialty hair and beauty products for cats and dogs.
- $3.4 million to Northeastern University in Boston by the NIH to have hamsters juiced with steroids and fight each other in cage matches to examine their "aggression and anxiety."

Hopefully bookkeeping blunders:

- $4.2 billion lost to improper tax refunds issued to identity thieves.
- $4 billion in funding to states who improperly achieved a double benefit on federal Medicaid payments.
- $120 million in retirement and disability benefits to federal employees who are dead.

Sci-Fi fantasy and foolishness:

- $80 million for the development of a real-life "Ironman" suit.

- $37 million for an initial inquiry as to the total cost for the United States to quell the rising unrest in the Middle East by "having everyone pretend to convert to Islam for a year or two."
- $1.2 million to study whether eating radioactive tuna caused by Japan's Fukushima disaster will provide humans with superpowers.
- $10,000 for talking urinal cakes that supposedly will discourage heavy drinking in bars.
- $2.6 million to encourage Chinese prostitutes to drink more responsibly.
- $442,340 to study male Vietnamese prostitutes by the National Institutes of Health (NIH).
- $5 million grant to Brown University from the NIH to study fraternities and sororities, which came to the shocking conclusion that students in fraternities and sororities consume more alcohol than other college students.
- $35,000 grant from National Highway Traffic Administration went toward a "Giant Marijuana Joint" billboard in Denver, Colorado, that can be seen in the dark as a means to discourage driving while stoned. Wow, man, check out the colors.
- $1.5 million from the NIH to Brigham and Women's Hospital in Boston to study why "three-quarters" of lesbians in the United States are overweight and why most gay males are not.
- $350,000 to the National Science Foundation for researchers at Purdue University to partially fund a study that investigated whether golfers' putting is improved if they imagine that the hole is bigger.
- $800,000 by the NIH to study the significance of a genital-washing program on South African men.
- $30 million to help Pakistani farmers grow mangos, basically a tropical fruit, in the desert.
- $325,525 for a study to investigate if the "happiest" marriages were the ones in which the wives were able to calm down quickly during conflict.
- $297 million and four years of the army's time and money for developing a mega-blimp that it eventually scrapped. But they sold the airship back to the contractor that built it for $301,000.

- $300,000 to determine who plays with Barbie dolls more, girls or boys. Most mothers could tell you that for nothing.
- $150,000 for a study to understand why politics stresses us out. (How about the waste and bloat of government spending).

Money madness:

- $12 billion in shrink-wrapped $100 bills was distributed in Iraq in 2004, and there was no control over who got it and how it was being spent.
- $100 million is spent every four years to subsidize parties at the political conventions.
- $79,000 was spent by the state department on booze for 10 American embassies for one year.

WHERE DOES THE MONEY AND REALITY GO?

Buried in the Department of the Treasury's *Financial Report of the United States Government* is a short section titled "Unreconciled Transactions Affecting the Change in Net Position," totaling $24.5 billion in 2003. The government knows that the money was spent by someone, somewhere, on something, but auditors do not know who spent it, where it was spent, or what it bought. The unreconciled billions could have funded the entire Department of Justice (DoJ) for an entire year.

In 2004, the Department of Defense (DoD) had unused flight tickets totaling $100 million, or about 270,000 fully refundable commercial airline tickets, for which they never sought refunds. Some employees even got reimbursed for the cost of a ticket, even though they never bought one. Another $140 million was wasted on upgrading from coach.

Credit cards issued to federal employees were designed to save money. But a recent audit revealed that employees of the USDA used the cards for personal items and 15 percent abused their government credit cards at a cost of $5.8 million. There are approximately 55,000 USDA credit cards in circulation, including 1,549 that are still held by people who no longer work there.

Over one recent 18-month period, air force and navy personnel used DoD credit cards to charge at least $102,400 for admission to entertainment events, $48,250 for gambling, $69,300 for cruises, and $73,950 for exotic dance clubs and prostitutes.

If anyone needed proof of how incredibly difficult it is to lose a government job, the DoJ found that Drug Enforcement Agency employees caught patronizing prostitutes were given bonuses rather than being fired, and the Government Accountability Office (GAO) reports that five federal agencies spent $3.1 billion on workers placed on administrative leave in a two-year time span. A lot of that cash—$775 million—went to public employees banned from their desks for more than a month for misconduct.

That's Entertainment

The National Science Foundation spent $3 million on a study that concluded the famous music in *Jaws* causes people to view sharks in a negative manner. How about $550,000 for *Rockin' the Kremlin*, a documentary on how rock and roll contributed to the end of the Cold War.

The National Endowment of the Arts awarded approximately $10,000 for a Christmas-themed show titled, "Mooseltoe: A New Moosical." The taxpayers produced "Zombie in Love," a musical about teenage undead "dying to find true love," for only $10,000. And for $15,000 the Colorado Symphony Orchestra produced "Classically Cannabis: The High Note Series," with the intention of attracting younger audiences to the symphony. Another $10,000 was paid out to restage "RoosevElvis," a play about a shy woman who channels the personalities of Elvis Presley and Teddy Roosevelt. And $27,000 to produce "One-Man Jurassic Park," a play meant to terrify and tantalize audiences with a tale of how "mankind's desire to play God backfires in spectacular fashion."

Our government spent $702,558 to bring television to Vietnamese villages in order to investigate how the tube affects family formation and reproductive health—even though the villages selected didn't have electricity, let alone TV sets. So the government paid to bring in TVs and gas generators. It didn't mention if couples made love during commercials.

Taxpayers shelled out $120,000 to Environmental Protection Agency employees who admitted to viewing porn up to six hours a day on government computers.

Public Funds for Programming

The public broadcasting service (PBS) is funded by the Corporation for Public Broadcasting (CPB), which gets infused with about $445 million dollars from the federal government. That may sound like a lot, but it costs each American only $1.37 in taxes per year, according to *Time* magazine. Over the course of a year, 80 percent of all U.S. television households, nearly 200 million people, watch PBS, a larger audience than most cable TV networks. National Public Radio (NPR) reaches 99 million people monthly and is funded by on-air pledge drives, corporate underwriting, state and local governments, and educational institutions. Federally funded CPB amounts to approximately 2 percent of NPR's overall revenues.

Getting Fat in Afghanistan

A 2011 DoD report found hundreds of defense contractors that used fraud and waste in dealings with the U.S. military, part of the cost of the business of war for the $1.1 trillion in Pentagon contracts between 2000 and 2010.

In 2014, a defense contractor responsible for providing food and water to troops in Afghanistan pleaded guilty to overcharging the U.S. government to the tune of $48 million. Two San Diego defense contractors pleaded guilty in a scheme that defrauded the navy out of at least $1.4 million by overbilling for supplies that the military never even ordered, the *San Diego Union-Tribune* reported.

Over the course of several years, the Defense Contract Audit Agency found that $553 million in payments should be disallowed to KBR, a Halliburton subsidiary, according to 2009 testimony by agency director April Stephenson before the bipartisan Commission on Wartime Contracting in Iraq and Afghanistan. When *Politifact*, a project that fact-checks statements by members of the government and special-interest groups, asked for a comment from KBR, a spokesperson replied, "Halliburton cannot comment on activity that relates to KBR's work in Iraq and Afghanistan as it would be inappropriate for Halliburton to comment on the merits of a matter affecting only the interest of KBR." No kidding, and a great example of double speak.

Hefty Health Bill

In 2012, Donald Berwick, a former head of the Centers for Medicare and Medicaid Services (CMS), and Andrew Hackbarth of the Rand Corporation, estimated that fraud added as much as $98 billion (roughly 10 percent), to annual Medicare and Medicaid spending and waste, and up to $272 billion (approximately 30 percent) across the entire health system.

It's not the poor and older people who need help with prescriptions or can't afford independent medical attention, and thus may fudge on their financials. Very few people are trying to rip off taxpayers for surgeries or bad teeth. Rather, it's the unscrupulous (or sometimes just incompetent) doctors and hospitals that provide unneeded goods and services and bloat their budgets with medical waste (see chapter 13, "Opening Pandora's Pharmacy").

Food Stamp Fraud—Frivolous or Sizeable?

When a family is having difficulty making ends meet, they often seek food stamps (about 14 percent of the population and more than one in five children are at risk of hunger in the United States). When someone intentionally provides misleading information about the size of their household or the amount of their income on an application for food stamps, they could be guilty of food stamp fraud.

Because of the economic downturn, from which many have not recovered, and because the supplemental nutrition assistance program (SNAP) has grown so exponentially with the economically disadvantaged, who don't have political clout, the program is a target for budget- and waste-minded lawmakers. They claim that many rip-off benefits for which they aren't eligible and sell off food stamps for cash or to buy nonfood items, such as booze, guns, or drugs. Actually, statistics show that only about 1.5 percent of assistance claims are fraudulent, according to *Time* magazine, and the epithet "welfare queen" that Ronnie Reagan exaggerated, politicized, and perpetuated was a fiction.

USE IT OR LOSE IT—FISCAL FEEDING FRENZY

Every September, the end of the fiscal year sparks a "use it or lose it" spastic spending seizure as federal agencies race to use up what's left in their annual budgets. Agencies often try to spend everything that's left instead of admitting they can operate on less because they are afraid that if they spend less than their budget allows, Congress might cut their budget next year.

President Trump was also in on this year's spending splurge. In the last week of fiscal year 2017, Trump's office spent $21.8 million, which is more than three times the $6 million former president Barack Obama's office spent to close out 2016. Trump's spending included $6.2 million in electrical hardware and supplies; $490,000 on tents and tarps; $489,517 on furniture; $10,612 on floor coverings; and $197,438 on newspapers and periodicals.

The Department of Health and Human Services spent more than $2 billion, including a $1.5 million deal with Square One Armoring Services for a fleet of armored vehicles.

The Department of Agriculture spent $306,617 on guns and an array of ammunition. In addition to guns and ammo, the agency loaded up on night-vision equipment ($1.5 million), personal armor ($3.5 million), and combat and tactical vehicles ($284,457). It must be hell protecting farming, agriculture, forestry, and food.

The federal government did some end-of-the-year cleaning, paying $152.5 million in "housekeeping" bills including janitorial services at $24.3 million; laundry and dry cleaning, $2.9 million; trash and garbage collection, $1 million; carpet cleaning, $630,943; and snow and ice removal, $127,373.

Many federal agencies decided to revamp and redecorate. In one week, the government spent $83.4 million on furniture plus another $23 million on office supplies and equipment. The Department of Veterans Affairs spent $15.6 million on new office furniture. The government also spent $862,000 annually to store more than 20,000 pieces of furniture that aren't being used. Craig's List anyone?

Fourteen agencies splurged with $3.5 million for clothes. The State Department and the Department of Homeland Security each spent more than $1 million on clothing, outerwear, and footwear in a one-week shopping spree.

Multiply this kind of free-for-all spending by hundreds of agencies and you'll understand why Washington's finances are such a mess.

Government Gets a Stay Out of Jail Free Pass

The irony is that a private company would quickly be out of business, and very soon in court, if it engaged in this kind of continual overspending, fraud, waste, and mismanagement found in government financial fiascos.

At home, we lock the doors and windows, install security systems, maybe have a watchdog, and neighborhood watch groups. Even though the government has some security systems and oversight committees, it doesn't seem to inhibit or stanch the green-back bleeding. Massive theft of the government's (i.e., taxpayers') money and property goes on and is tolerated decade after decade because too few in government care "about locking the doors." They all claim, "That's not my job."

Congress is made up of 535 people, part of whose business it is to judiciously spend trillions of the taxpayers' dollars. Americans should express righteous anger about the details in the annual reports of spending idiocy and abuse. Some do, but it tends to be forgotten by the next election cycle, as people usually vote for the incumbent, who probably had an open hand in the spending game in the first place.

How come Washington doesn't have agencies to monitor and expose government waste and mismanagement of our tax dollars? We do. There are 73 inspectors general at different federal agencies. The White House contains an Office of Management and Budget that looks for ways to cut costs. The GAO also checks expenditures. And so do legions of congressional staffers who are eager to help their bosses score political points. Over the past three years, the GAO found 162 areas where agencies are duplicating efforts, at a cost of tens of billions of dollars.

A Plethora of Redundant Programs

The GAO estimated the government could save up to $200 billion over the next decade by consolidating some of:

- 342 economic development programs

- 130 programs serving the disabled
- 130 programs serving at-risk youth
- 90 early childhood development programs
- 75 programs funding international education, cultural, and training exchange activities
- 72 federal programs dedicated to assuring safe water
- 50 homeless assistance programs
- 40 separate employment and training programs
- 27 teen pregnancy programs
- 26 K–12 school grant programs
- 23 agencies providing aid to the former Soviet Republics
- 19 programs fighting substance abuse
- 17 rural water and waste-water programs in eight agencies
- 17 trade agencies monitoring 400 international trade agreements
- 12 food safety agencies
- 13 agencies spend a total of $30 million annually funding 15 separate financial literacy programs
- 11 federal agencies operate 94 separate initiatives to spur energy-efficient construction in the private sector
- 11 statistics agencies
- 9 agencies' or departments' programs to safeguard food and agricultural systems from natural disasters and terrorist attacks

You think there might be a little overlap or redundancies with 1,114 separate programs or agencies?

"A billion here, a billion there. Pretty soon you're talking about real money," Senator Everett Dirksen said fifty years ago. But those itty bitty billions are bloating the $4 trillion that make up the federal budget. And yet the pork trough fills up every year. It makes one wonder what's going to happen 50 years from now.

Taxpayers should think of themselves as "investors" in the U.S. government—the largest financial entity in the world. Would you hire a CEO of a company with a budget of trillions a year who had no management training, financial education, and common sense? Then again, we elected a president with no political or foreign affairs experience.

GOVERNMENT WASTE REDUX

Politicians have been promising to win the war on fraud, waste, and abuse of tax money for as long as we've had voters. But the disappointing outcomes suggest they still haven't found a way that works. Back in 1949, President Truman directed ex-president Herbert Hoover to organize 300 men and women to seek and eliminate waste in government. Ronald Reagan, in his first inaugural address in 1981, vowed "to curb the size and influence of the federal establishment," because, "the federal government is not part of the solution, but part of the problem." In 1982, Reagan asked investigators to "work like tireless bloodhounds" to "root out inefficiency, and take out the waste and fraud that was eroding faith in our government." In 2012, President Barack Obama asked Vice President Joe Biden to spearhead a "campaign to cut waste" and restore citizen trust in government. President Trump signed an executive order aimed at cutting waste in the federal government in 2017.

Yet despite decades of pledges, campaigns, reports, and committees, the challenge and promises to our citizens remain unfulfilled, and the pork barrels remain filled to the brim. After the 2016 election, the House Republican Conference failed to renew the earmark moratorium for the first time since it was first adopted. And recently, President Trump piped in with, "I hear so much about the old earmark system, how there was a great friendliness when you had earmarks." In other words, it's true that quid pro quo earmarks make it easier to grease the wheel with pork fat and pass bills—which is exactly the problem.

"If there is one thing I learned during my time in Congress, it is never to underestimate the dumb things politicians will dream up to spend other people's money on," said ex-representative Tom Coburn.

SOURCES

"11 Facts about Hunger in the US." Do Something, https://www.dosomething.org/facts/11-facts-about-hunger-usa.

"30 Stupid Things the Gov't Spends Money On." Space Battles.com, March 4, 2012, https://forums.spacebattles.com/threads/30-stupid-things-the-govt-spends-money-on.218209/.

Bandler, Aaron. "9 Ridiculous Things the Government Wasted Money on This Year." *Daily Wire*, January 11, 2017, https://www.dailywire.com/news/12309/9-ridiculous-things-government-wasted-money-year-aaron-bandler.

Adams, Becket. "Here Are the Top Six Most Ridiculous Things the Gov't Spends Tax Dollars On." *The Blaze*, December 17, 2013, https://www.theblaze.com/news/2013/12/17/here-are-the-top-6-most-ridiculous-things-the-govt-spends-tax-dollars-on.

Andrzejewski, Adam. "Use It or Lose It: Trump's Agencies Spent $11 Billion Last Week in Year-End Spending Spree." *Forbes*, October 3, 2017, https://www.forbes.com/sites/adamandrzejewski/2017/10/03/use-it-or-lose-it-the-federal-governments-11-billion-year-end-spending-spree/#6e76f5c67697.

Axe, Jack. "Pentagon Weapons-Buying: 'Dumb as a Bag of $600 Hammers.'" *Wired*, September 19, 2008, https://www.wired.com/2008/09/dumb-as-a-bag-o/.

Bruce, Kalen. "24 Stupidest Things the U.S. Government Spends Money On." *MoneyMiniblog*, June 29, 2015, http://moneyminiblog.com/interesting/stupidest-things-u-s-government-spends-money-on/.

Calder, Vanessa. "Why Welfare Needs Reform." Cato Institute, January 22, 2018, https://www.cato.org/publications/commentary/why-welfare-needs-reform.

Cobern, Tom. "Earmarks Are Inherently Corrupt Congress Has No Business Resurrecting Pork Barrel Politics." *Federalist*, January 11, 2018, /http://thefederalist.com/2018/01/11/earmarks-are-inherently-corrupt-congress-shouldnt-resurrect-them/.

"Corruption, Abramoff, the Man and the Mentality." *Public Citizen Watchdog*, June 27, 2006, http://citizen.typepad.com/watchdog_blog/ethics/page/2/.

Dart, Andrew. "Specific Examples of Pork Barrel Spending." http://www.akdart.com/pork3.html.

Davis, Christina. "San Diego Contractors Admit Overbilling Navy $1.4 Million in Fraudulent Supply Orders." *San Diego Tribune*, February 28, 2017, http://www.sandiegouniontribune.com/news/courts/sd-me-contractor-fraud-20170228-story.html.

Angie Drobnic, Holan. "Halliburton, KBR, and Iraq War Contracting: A History so Far." *Politifact*, June 9, 2010, http://www.politifact.com/truth-o-meter/statements/2010/jun/09/arianna-huffington/halliburton-kbr-and-iraq-war-contracting-history-s/.

Deppin, Colin. "Pa. Defense Contractor Bilked U.S. Gov't Out of $6 million for Humvee Parts, DOJ Alleges." *PennLive*, March 5, 2017, http://www.pennlive.com/news/2017/03/pa_defense_contractor_bilked_u.html.

Dicker, Rachel. "Paul's Festivus Reveals $1B in Wasteful Government Spending." *U.S. News and World Report*, December 24, 2015, https://www.usnews.com/news/slideshows/rand-pauls-festivus-reveals-1b-in-wasteful-government-spending.

Doherty, Daniel. "Sen. Coburn's 'Wastebook 2014': $10,000 Spent on 'Watching Grass Grow' & Other Crazy Things." *Town Hall*, October 22, 2014, https://townhall.com/tipsheet/danieldoherty/2014/10/22/sen-coburns-wastebook-2014-n1908529.

Fabian, Jordan. "Trump Signs Executive Order to Cut Government Waste." *The Hill*, http://thehill.com/homenews/administration/323772-trump-signs-executive-order-to-cut-government-waste.

Fine, Glen. "Seven Principles of Highly Effective Inspectors General." Center of Advancement of Public Integrity, January 2016, https://media.defense.gov/2017/Nov/16/2001844771/-1/-1/1/SEVEN_PRINCIPLES%20_OF_HIGHLY_EFFECTIVE_INSPECTORS_GENERAL.PDF.

"Everett Dirksen." Wikiquote, https://en.wikiquote.org/wiki/Everett_Dirksen.

"Politicians Routinely Lie and Act for Their Own Interests, Often Determined by Pressure Groups, Instead of the Peoples' Interest." Government Does More Harm Than Good. https://sites.google.com/site/governmentdoesmoreharmthangood/3-politicians-routinely-lie-and-act-for-their-own-interests-often-determined-by-pressure-groups-instead-of-the-peoples.

"Government Waste by the Numbers: Report Identifies Dozens of Overlapping Programs." *Fox News*, March 1, 2011, http://www.foxnews.com/politics/2011/03/01/government-waste-numbers-report-identifies-dozens-duplicative-programs.html.

Hoel, Leland. "Hidden Entitlements." *My Mixed Blog*, July 30, 2017, https://lelandolson.com/2017/07/30/11483/.

"How the U.S. Sent $12bn in Cash to Iraq. And Watched It Vanish." *Guardian*, https://www.theguardian.com/world/2007/feb/08/usa.iraq1.

Henderson, Sara. "What Is Article 1 Section 8 of the U.S. Constitution?" *Pocket Sense*, https://pocketsense.com/article-1-section-8-us-constitution-19300.html.

Hensarling, Jeb. "Pork-Barrel Spending." *I Spy*, May 14, 2008, https://votesmart.org/public-statement/344444/pork-barrel-spending#.WujHxjNlDcs.

"Jeff Flake's Full Speech from the Senate Floor." *CNBC.com*, https://video.search.yahoo.com/yhs/search?fr=yhs-iry-fullyhosted_003&hsimp=yhs-fullyhosted_003&hspart=iry&p=Arizona+Senator+Jeff+Flake+%E2%80%9Cyou+can%E2%80%99t+drain+the+swamps+by+feeding+the+alligators+pork.%E2%80%9D#id=1&vid=d6a85fb15368d65ce377ff4c34e52e28&action=click.

Krieger, Michael. "'Draining the Swamp'—Trump Admin Blows $11 Billion in the Last Week of Fiscal Year." *Liberty Blitzkrieg*, October 5, 2017, https://libertyblitzkrieg.com/2017/10/05/draining-the-swamp-trump-admin-blows-11-billion-in-the-last-week-of-fiscal-year/.

Krist, Kathy. "10 Most Outrageous Ways Government Wastes Your Money." *Money Watch*, December 20, 2011, https://www.cbsnews.com/news/10-most-outrageous-ways-government-wastes-your-money/.

Korte, Gregory. "Report: Redundant Federal Programs Waste Billions." *USA Today*, April 9, 2013, https://www.usatoday.com/story/news/politics/2013/04/09/wasteful-government-spending/2063511/.

"Levin, Josh, "The Welfare Queen." *Slate*, December 19, 2013, http://www.slate.com/articles/news_and_politics/history/2013/12/linda_taylor_welfare_queen_ronald_reagan_made_her_a_notorious_american_villain.html.

Mitchell, Dan. "Government Fraud: A Feature, Not a Bug." *International Liberty*, July 9, 2017, https://danieljmitchell.wordpress.com/2017/07/09/government-fraud-a-feature-not-a-bug/.

Martosko, David. "How Much Pornography Would It Take for an EPA Employee to Lose Their Job?" *Daily Mail*, May 7, 2014, http://www.dailymail.co.uk/news/article-2622503/How-pornography-EPA-employee-lose-job-Congress-fumes-daily-porn-surfing-EPA-employee-STILL-collecting-120-000-salary.html.

"McCain Leading Charge against Earmark-Stuffed Spending Bill." *Fox News*, December 15, 2010.

"Monkeys Gambling with Your Money." *2014 Wastebook*, https://www.scribd.com/document/243970542/Wastebook-2014.

Neff, Blake. "9 Shockingly Stupid Examples of Federal Government Waste." *Daily Caller*, November 29, 2016, http://dailycaller.com/2016/11/29/smell-museums-feminist-glaciers-and-7-other-embarrassing-cases-of-federal-waste/.

Nitty, Tony. "How Did the Government Waste Your Tax Dollars in 2014? Separating Fact from Fiction." *Forbes*, October 22, 2014, https://www.forbes.com/sites/anthonynitti/2014/10/22/how-did-the-government-waste-your-money-in-2014-separating-fact-from-fiction/#1f7cf2a027ca.

"The Favor Factory." *Washington Post*, July 31, 2006, http://www.washingtonpost.com/wp-dyn/content/article/2006/07/30/AR2006073000540.html.

Offensicht, Will. "Yes, Virginia, A $2.98 Hammer REALLY Costs Our Government $100." *Scragged*, January 15, 2010, http://www.scragged.com/articles/yes-virginia-a-$2.98-hammer-really-costs-our-government-100.

"PBS Funding Standards." KQED, http://www.pbs.org/about/producing-pbs/funding/.

Papst, Chris. "Department of Justice Indicts U.S. Congressman on Corruption Charges." WJLA, July 29, 2015, http://wjla.com/news/crime/department-of-justice-indicts-us-congressman-on-corruption-charges-07-29-2015.

"Pork Barrel." Wikipedia, https://en.wikipedia.org/wiki/Pork_barrel.

Rainey, Michael. "The Pentagon's $65 Million Plane That Never Flew a Mission." *Fiscal Times*, September 29, 2017, http://www.thefiscaltimes.com/2017/09/29/Pentagon-s-65-Million-Plane-Never-Flew-Mission.

Rappeport, Alan. "To Grease Wheels of Congress, Trump Suggests Bringing Back Pork." *New York Times*, January 10, 2018, https://www.nytimes.com/2018/01/10/us/politics/trump-earmarks-pork-barrel-spending.html.

Riedl, Brian. "Top 10 Examples of Government Waste." The Heritage Foundation, April 4, 2005, https://www.heritage.org/budget-and-spending/report/top-10-examples-government-waste.

"Ronald Reagan, First Inaugural Address." Bartleby.com, http://www.bartleby.com/124/pres61.html.

Rude, Emelyn, "The Very Short History of Food Stamp Fraud in America" *Time*, March 30, 2017, https://www.yahoo.com/news/very-short-history-food-stamp-150008832.html.

Sanibel, Michael. "Money and Politics." *Investopedia*, May 3, 2012, https://www.yahoo.com/news/money-politics-205534712.html.

Schatz, Tom. "Congress Must Rid Itself of Political 'Pork' to Preserve Its Integrity." *The Hill*, July 12, 2017, http://thehill.com/blogs/pundits-blog/lawmaker-news/341704-congress-must-rid-itself-of-the-dirty-politics-of-pork.

Scudde, Casey. "The 10 Most Absurd Pork Barrel Spending Items of 2010." *Nasdaq*, August 13, 2010, https://www.nasdaq.com/article/the-10-most-absurd-pork-barrel-spending-items-of-2010-cm32756.

Smith, Jack. "$37 Screws, a $7,622 Coffee Maker, $640 Toilet Seats: Suppliers to Our Military Just Won't Be Oversold." *Los Angeles Times*, July 30, 1986, http://articles.latimes.com/1986-07-30/news/vw-18804_1_nut.

Schnurer, Eric. "Just How Wrong Is Conventional Wisdom about Government Fraud?" *Atlantic*, August 15, 2013, https://www.theatlantic.com/politics/archive/2013/08/just-how-wrong-is-conventional-wisdom-about-government-fraud/278690/.

Snyder, Michael. "30 Stupid Things the Government Is Spending Money On." *American Dream*, February 29, 2012, http://endoftheamericandream.com/archives/30-stupid-things-the-government-is-spending-money-on.

Steel, Michael. "Commentary: Congress Is Broken, but Earmarks Will Only Make It Worse." *The Bulletin*, January 21, 2018, http://www.bendbulletin.com/opinion/5930018-151/commentary-congress-is-broken-but-earmarks-will-only.

"Tammany Hall." Wikipedia, https://en.wikipedia.org/wiki/Tammany_Hall.

"The $272 Billion Swindle, Why Thieves Love America's Health-Care System." *Economist*, May 31, 2104, https://www.economist.com/news/united-states/21603078-why-thieves-love-americas-health-care-system-272-billion-swindle.

"The Grace Commission." https://www.revolvy.com/topic/The%20Grace%20Commission.

"The Top 10 Examples of Government Waste." *Democratic Underground.com*, 2005, https://www.democraticunderground.com/discuss/duboard.php?az=view_all&address=104x3430118.

"United States Congress." Wikipedia, https://en.wikipedia.org/wiki/United_States_Congress.

Utt, David. "The Bridge to Nowhere: A National Embarrassment." The Heritage Foundation, October 20, 2005, https://www.heritage.org/budget-and-spending/report/the-bridge-nowhere-national-embarrassment.

Viechnicki, Peter, et al. "Shutting Down Fraud, Waste, and Abuse: Moving from Rhetoric to Real Solutions in Government Benefit Programs." *Deloitte Insights*, May 11, 2016, https://www2.deloitte.com/insights/us/en/industry/public-sector/fraud-waste-and-abuse-in-entitlement-programs-benefits-fraud.html.

Warner, Joel. "Puff Puff Brass." *Slate*, May 29, 2014, http://www.slate.com/articles/arts/culturebox/2014/05/classically_cannabis_concert_review_colorado_symphony_orchestra_s_high_note.html.

"Wasteful Spending List." U.S. Congressman Bill Posey, https://posey.house.gov/wasteful-spending/.

"What Are Some Examples of "Pork Barrel Politics" in the United States?" *Investopedia*, https://www.investopedia.com/ask/answers/042115/what-are-some-examples-pork-barrel-politics-united-states.asp.

"Where Did the Phrase "Pork Barrel" Come From?" *Investopedia*, https://www.investopedia.com/ask/answers/050615/where-did-phrase-pork-barrel-come.asp.

17

THE FORTY-HOUR WORKWEEK WASTE

> One of the biggest failures of business . . . is that we are measured by how much we work and not by what we accomplish.
> —*Chris Bailey*

It's the wheel that grinds the grist for the American economy—the 40-hour workweek. The Baby Boomers and their parents grew up to expect and respect it, but the Gen Xers and Gen Ys have different ideas about the traditional workweek, and about how much time to spend there.

Let's not kid ourselves, everyone at some time or another has goofed off at work, whether it's using company time or tools for personal use, shaving a little time off to surf the net for politics, games, shopping, or porn, chatting with friends and family, or taking a not-really-sick day. Eighty percent of employees waste time on the clock, according to Getvoip, which offers comparison guides and rankings to assist businesses.

There's talk now about whether the 40-hour workweek is a Victorian relic that has outlived its need and effectiveness. There are certainly a lot of alternatives—flextime, part-time, work at home, shared responsibilities, partial retirement, etc. So maybe it's time for the workweek to get a makeover, or at least be modified to a more sustainable model of efficiency. In fact, it's one perk that most workers desire.

Back in the day, people worked from sun to sun, hard work was part of our expected heritage. Things didn't change until 1869, when President Ulysses S. Grant issued a proclamation that ensured that all government workers would work eight hours per day. In 1916, President Woodrow Wilson made the case for giving railroad workers an eight-hour day when railroad unions pointed out that longer hours caused a rise in accidents and death. The next big break came from Henry Ford in 1926, when he instituted the 40-hour workweek in his factories. It was not from conscience or empathy, Ford wanted his employees to have time to spend their money on his cars. When President Franklin D. Roosevelt implemented his New Deal Reforms, one of them was the eight-hour day. And in 1940 Congress amended the Fair Labor Standards Act, setting the workweek to a maximum of 40 hours, still the standard today.

Americans are working more than ever; 58 percent of managers in the United States reported working more than 40 hours a week, and the average for full-time workers is 47 hours. A Gallup Poll of 1,200 American adults, found that 18 percent worked 60 hours or more every week, with another 21 percent claiming to work between 50 and 59 hours. Still another 11 percent estimated between 41 and 49 hours per week. Occasionally, this is due to personal motivation, but more often it's due to pressure from upper management or peers to stay later and work harder.

In many ways, the 40-hour workweek is already dead. In some cases, the regimented schedule is completely ignored in favor of meeting certain project needs. In others, it is so strictly adhered to that it becomes a leftover of an earlier era. Either way, the pretense of a standardized 40-hour workweek is an obsolete notion, and more companies are looking at alternatives.

JUST BECAUSE YOU'RE AT WORK DOESN'T MEAN YOU'RE WORKING

You're drowning in email, stuck in dead-end meetings, and constantly interrupted. A recent survey indicated that American workers spend up to 6.3 hours every day checking and managing email; that's 16 times a day, and then it takes 10 to 20 minutes spent refocusing. It's the same as missing an entire night's sleep. Even if half that time is spent on personal emails, that's still 15 hours per week spent managing work-related emails.

Although 26 percent said the biggest overall time-wasting activity was browsing the Internet, the old standbys were right behind. Returning emails garnered 12 percent, dealing with an annoying boss was 7 percent, social media was at 4 percent, and personal phone calls were 2 percent of respondents.

Time spent with coworkers is not always time spent getting work done. Pointless and multiple meetings, conference calls, chatting, and dealing with annoying coworkers were the worst time wasters. The bottom line is, according to Basex research, interruptions at work cost the U.S. economy $588 billion a year.

THE DUES AND DON'TS OF OVERWORK

Studies by Marianna Virtanen of the Finnish Institute of Occupational Health and her colleagues have found that overwork and the resulting stress can lead to all sorts of health problems, including impaired sleep, depression, heavy drinking, diabetes, impaired memory, heart disease, impairment of communication, and judgment calls. Not only are those hard on an employee, they're also terrible for a company's bottom line, and rising health insurance costs.

Simply stated, anyone is more likely to make mistakes when tired. Just 1 to 3 percent of the population can sleep only five or six hours a night without suffering some performance loss.

In fact, the *Annals of Internal Medicine* studies even claim that long hours can literally make you crazy. Working more than 39 hours a week can wreak havoc on your mental health and cause other serious sicknesses, according to research published in the journal *Social Science & Medicine*. To stay healthy and productive, women should work no more than 34 hours per week, according to the researchers. Women—who are often slammed with tasks both at home and the workplace—are more prone to stress, researchers said. The limit was higher for men, who researchers capped at 47 hours a week.

Spending work time doing personal stuff online is an increasingly common habit that could be even more addictive than coffee, according to new research into Internet usage in the office. According a Web@Work survey, 93 percent of all employees in the United States spend at least some time at work accessing the Web, up from 86 percent a year ago. Among those, 52 percent said they would rather give up their morning caffeine hit than lose their Internet connection.

The average time spent accessing the Internet at work was 12.6 hours per week. While employees estimated that 3.4 hours of that time was due to nonwork-related surfing, managers put that figure at closer to six hours. Websense, a computer information firm, claims that this behavior costs American corporations nearly $200 billion per year.

The Bummer of Burnout and Dazed Days

The Centers for Disease Control and Prevention cites studies that found "a pattern of deteriorating performance on physiological tests as well as injuries while working long hours." It also cited four studies that found that

> by the eighth hour of the day, people's best work is usually already behind them, as fatigue sets in and they're only going to deliver a fraction of their efficiency. And with every extra hour beyond that, the workers' productivity level continues to drop. The 9th to 12th hours of work were tainted with feelings of decreased alertness and increased fatigue, lower effective thinking, and at around 10 or 12 hours they hit exhaustion.

Friday afternoons are still the prime time to waste time. Forty-four percent said they waste the most time on Fridays, and 22 percent said that they waste the most time between 3 p.m. and 5 p.m. Next is Monday afternoon, with 18 percent choosing 1 p.m. to 3 p.m.

Incidentally, our brains aren't wired to concentrate intensely for eight hours straight. Logging more hours doesn't necessarily add up to more productivity. According to project management software Podio, your brain can only focus for 90 to 120 minutes, at which point it needs a short break before you can launch into your next period of focus. And by the time 3 p.m. rolls around, you're working least effectively—better to go home and rest.

EVERYONE WASTES TIME AT WORK

According to a Web survey by America Online and Salary.com, the average worker admits to frittering away 2.09 hours per day, not counting lunch. Over the course of a year, that adds up to $759 billion in salaries for which companies receive no apparent benefit.

The amount of people wasting three or more hours at work was surprisingly high at 18.5 percent. In a year, it adds up to 780 hours, or nearly 98 wasted workdays. Not only does this hamper employees' output, it also means that nearly one in five employees are getting paid for a lot of hours that they aren't actually working.

To put this in perspective dollar wise, the average American worker makes $49,094 a year, or $25.57 an hour. If we make a conservative estimate and assume most of these people waste three hours per day and are paid an average salary, we see that employers are spending $340.65 a week, or $18,401.04 a year, per employee on wasted time.

Time Wasting Rationalizations

Paychex, an American provider of human resource and outsourcing services, asked 2,000 workers to come clean about what prompts them to goof off at work. Amazingly, over 20 percent of their respondents say a lack of work is the primary reason for not working, while almost 15 percent say it's because they're dissatisfied or bored at work. Nearly 14 percent say their long work hours prompt them to waste time, and more than 11 percent say they're simply not motivated. Only 7 percent claimed wasted time was due to a coworker or a boss issue, about 3 percent do so because of low pay, and fewer than 2 percent due to a lack of time off. Nearly one-fifth of respondents confessed they're easily distracted, and almost 7 percent say they're sleep deprived or stressed. It looks like many employees might be motivated if they had more meaningful work, greater challenges, stronger incentives, and shorter work hours.

Some actually claim they think taking "time off" is proactive. And they might be right. Of all those surveyed by Salary.com, 53 percent say they waste time because they believe short breaks actually increase productivity. The survey collected data from 750 employees, 89 percent of which admitted to wasting time in the workplace—a dramatic increase from 69 percent the previous year.

Another survey, by Harris Poll for CareerBuilder, questioned 2,138 hiring and human resource managers and 3,022 full-time employees from a wide range of fields and company sizes to determine the causes of wasting time. As many as 24 percent of respondents confessed to spending at least an hour a day on personal email, texts, and calls. The list of time-wasting activities also included browsing social media sites, the Internet, taking snack or smoke breaks, meetings, and distractions by coworker visits and calls.

Thirteen percent of people surveyed said they fully intend to waste time watching games that take place during work hours. College basketball's March Madness, costs an estimated $175 million in wasted time in just the first couple of days—an estimated overall cost to the Gross National Product (GNP) of $1 billion.

Rationale for time out according to the poll:

- 34 percent say they are not challenged
- 34 percent say they work long hours
- 30 percent are unsatisfied with work
- 24 percent due to low wages
- 23 percent are bored
- 22.5 percent lack of work
- 21 percent claim no incentive
- 19.1 easily distracted at work
- 13.5 complain of long hours
- 6.5 percent stressed or sleep deprived
- 2 percent lack of time off

The Daily Waste

Many unhappy and nonmotivated employees simply plod through their work unfocused and get little done each day, and menial tasks become accepted as a way to fill time. Salary.com surveyed 2,063 adults between the ages of 25 and 64 and asked them how many hours they spent on nonwork activities during an average workday:

- 19.6 percent said they didn't waste any time

- 19.3 percent wasted an hour
- 18.5 percent wasted three or more hours
- 11.6 percent wasted two hour
- 8.6 percent wasted an hour and a half
- 4.6 percent wasted two and a half hours

How Workers Claim They Waste Time

Several years ago, 69 percent of respondents from a survey conducted by Salary.com said that they waste at least some time at work on a daily basis. But a more current and realistic survey has the number of people who admit to wasting time at work every day at an eye-popping 89 percent.

The CareerBuilder poll asked hiring managers and Human Resources (HR) professionals and full-time workers across a variety of industries and company sizes to determine how workers waste time. According to the survey:

- 50 percent talking on the cell phone and texting
- 42 percent gossiping
- 39 percent on the Internet
- 38 percent on social media
- 27 percent taking snack or smoke breaks
- 24 percent personal email, texting, personal calls
- 24 percent distracted by noisy coworkers
- 23 percent in meaningless meetings
- 23 percent on email

How bosses think workers waste time:

- 55 percent cell phone and texting
- 41 percent Internet
- 39 percent gossip
- 37 percent social media
- 27 percent coworkers stopping by
- 27 percent smoke or snack breaks
- 26 percent email
- 24 percent meetings

- 20 percent noisy coworkers
- 9 percent just sitting alone in a cubicle

Boss's Goof Time

There's a chance your boss is goofing off for more than a third of the workday. According to a study conducted by Paychex, approximately one in every 10 middle and upper manager wastes at least three or more hours a day. The study also revealed that as education level increases, an individual's level of focus decreases.

THE GENERATION GAPS

Employees can be sorted into three general categories: Baby Boomers, Generation X, and Generation Y (Millennials), and each wastes time in different ways for various reasons.

The Baby Boomers, born from 1946 to 1964, had no Internet or cell phones in college and are immigrants to a technology-based society, waste the least amount of time, and are more attuned to their parents' work ethic of eight hours of work or more.

Generation Xs were born between the years of 1965 and 1981; they had a mix of Internet and cell phones in their later years. They rank workplace flexibility as the most important perk and are more likely to walk away from their current job if flexibility isn't available. They are working in the same career as when they entered the workforce, and almost 25 percent have been with the same employer for 15 years or longer. The majority agree with the statement "hard work is the key to getting ahead" and 70 percent prefer to work independently.

The Millennials (or Gen Y) were born between 1982 and 2004 and had access to the Internet and cell phones while in school. Fifty-three percent say they would give up their sense of smell rather than lose their electronic connections. One-third of them would rather have a flexible work environment and access to social media than a bigger paycheck. They waste more than twice as much time as Boomers. On average, this tech-obsessed generation uses their devices 7.5 hours per day. They have developed an idea of being entitled to "me" time at work, and are highly team oriented, which naturally sets the stage for more conversation with coworkers, which leads to wasted time. The number one cause for distraction among Millennials is the Internet, upon which they spend an average of two hours per day at work, according to globalwebindex.net, a company that provides profiling data.

Young vs. Old, Men vs. Women

It depends on who does the survey of Internet time at work. It seems that men waste slightly more time than women at work—91 percent to 87 percent, with single men in their 20s and 30s without higher education wasting the most time. However, male employees spend 40 minutes more time in the office than female colleagues. When it comes to social media at work, women are at 27 percent use, men at 17 percent. The following statistics show that the older you are, the less time you waste at work:

- Baby Boomers, 41 minutes
- Gen Xers, 1.6 hours
- Millennials, 2 hours

CYBERSLACKING

We have at our fingertips the most powerful and the most seductive tool of the automated age—the World Wide Web via the Internet. What we call the social network is a massive automated collective that makes up 7 percent of the websites accessed at work and is a major contributor to all the wasted time in the workplace.

We now have access to more than two zettabytes of electronic fodder on which to forage, and personal use of technology continues to lead surfing and goofing off. (One zettabyte, or two trillion gigabytes is the equivalent of 36,000 years of high-definition video.)

The distractions of getting caught up in the net are endless, as volumes of new data and photos are uploaded instantaneously and constantly. Web surfers are bombarded with millions of fresh articles, emails, graphics, tweets, Instagrams, YouTube videos, and Snapchats every day. More than 1.1 billion active Facebook users upload 350 million photos daily, and more than 100 hours of video join the YouTube database every minute.

When workers become lackadaisical, shuffle back to their desks after an extended water-cooler conversation or long lunch, toggle between emails, scan to "like" a new picture or respond to an email or Instagram, "tweet," or check their status with professionals and friends on LinkedIn, these social media zombies are wasting big time on the greatest productivity drain of all time.

According to IDC Research, a provider of market intelligence, 30 to 40 percent of employee Internet activity is nonwork-related:

- Workplace Internet misuse costs U.S. businesses more than $63 billion in lost productivity annually, according to Websense Inc.
- Charles Schwab reveals that 72 percent of its customers plan to buy or sell mutual funds in the near future and 92 percent of these plan to do so online during work hours.
- 70 percent of all Internet porn traffic occurs during the 9 to 5 workday, according to SexTracker.
- 28 percent of individuals that make gift purchases do so from the workplace, according to Pew Internet and American Life Project.

According to recent research from Nielsen/NetRatings, fewer than 6 percent of Americans with Internet connections have high-speed access at home, and accessing new technology media takes a high-speed connection, found most frequently in corporate environments—where it's free to employees using corporate resources.

Napster music-swapping software (now defunct) was found on 20 percent of more than 15,000 work computers examined, according to eMarketer.com. Still contributing percentages of work time lost are Tumblr, 57 percent; Facebook, 52 percent; Twitter, 17 percent; LinkedIn, 14 percent; Instagram, 11 percent; Yahoo,7 percent; SnapChat, 4 percent; Amazon, 2 percent; YouTube, 2 percent; ESPN, 2 percent, and Pinterest and Craigslist each at 1 percent.

Electronic Missive Misuse

More than 70 percent of workers admit to checking their personal emails at work, 33 percent of them check it three times a day. Abuse is rampant—28 percent of employers have already fired workers for either violation of workplace policy and/or excessive personal use, and 6 percent for having used email to transmit customers' confidential information.

Thirty-three percent of workers' time is spent addressing emails that aren't urgent, or don't necessarily warrant a response. Yearly, $1,800 per user is lost reading and responding to unnecessary emails from coworkers, and $2,100 to $4,100 a year per user is wasted due to dealing with poorly written communications.

MORIBUND MEETINGS

One of the biggest beefs of employees is wasting time attending meetings that don't start on time, are underprepared, and lack leadership. Meetings that go off-course, where there is no clear purpose or objective, are boring and provide no new or important information, and nothing gets done afterward according to GiveMore.com. Almost half of all meetings are seen as a time-suck.

LEND ME YOUR EAR

According to a new survey from CareerBuilder of 2,186 hiring managers and 3,031 full-time employees, 20 percent of employers think that employees are productive fewer than five hours a day. Fifty-five percent said workers' cell phones are the biggest source of distraction. However, only 10 percent of respondents with smartphones said it decreases their productivity while at the office, yet 66 percent said they use their smartphones several times a day while working. And 82 percent of employees said they always keep their smartphone within reach or eye contact.

Furthermore, almost half of employers believed smartphone distractions compromised the quality of work, almost 30 percent claimed a negative impact on the boss–employee relationship and caused 28 percent missed deadlines. People check their phones for:

- 65 percent personal messaging

- 51 percent weather sites
- 44 percent news
- 24 percent games
- 24 percent shopping
- 12 percent traffic sites
- 7 percent gossip
- 6 percent sales
- 4 percent adult sites
- 3 percent dating sites

HEY WAKE UP—BIG BROTHER MAY BE SNOOPING

Getting busted for using work time for gossip and griping about the company, its clients, and management is easier than people think. Not only do these activities result in negative job reviews, they can be career busters. This knowledge should be an incentive not to leave electronic fingerprints behind.

Less than 20 percent of businesses have a social media policy. This means a lot of businesses are still losing money year after year, to the tune of more than $200 billion per year. Forty-five percent of employers are tracking computer content using network software, 43 percent also monitor email, 43 percent even store and review computer files, and roughly half of them actually assign someone to manually read and review emails sent on company time. And to top that, they're blocking websites at work. Facebook is the most blocked, followed by MySpace, YouTube, Doubleclick.net, and Twitter. While this may seem rather heavy-handed, 70 percent of employees say that employers have the right to monitor and/or block certain Internet activity.

SOME SOLUTIONS AND SUGGESTIONS

To limit distractions, some employers block certain sites, prohibit personal calls and cell phones, monitor emails, reduce meetings, discourage nonwork Internet use, and organize the office in an open space layout. But if it's not possible to structure your office this way, there are still many strategies that you can suggest to help employees stay focused.

- Set concrete goals of what you are going to get done for the day and allot blocks of time to them. This will help you focus on microtasks instead of feeling overwhelmed by all that you have to get done.
- Turn your phone on silent, don't use iMessage on your computer, and turn off notifications for personal emails to help minimize distractions.
- Listen to music or white noise to drown out noisy coworkers and help you concentrate.
- Create a shared calendar so people know your schedule and can determine the best time to come ask you questions or collaborate.
- Maintain a clean and organized workspace. This will help focus on what you're working on instead of getting distracted by feeling overwhelmed by a messy work area.

Some Alternatives

Rather than five eight-hour days, some companies are trying four 10-hour days or three 12-hour days. It's still a structured schedule, but it's a semiviable alternative. Other alternatives attractive to employees are flex-time where employees schedule their own hours, whether at the workplace or at home. Implementing this requires a high level of trust in your workers to get things done and work responsibly, but if you trust your workforce and your employees are passionate about what they do, the potential benefits could be remarkable.

Working from home has been a controversial employee perk for some time, but many employers swear by it. When working from home, employees are free from the distractions of the office, and they can focus on tasks with more autonomy and flexibility. On the other hand, it's also a good way to get a tan.

To keep employees awake and aware, some employers offer facilities for those who wish to exercise and refresh with a shower. That might mean offering discounts or subsidies on memberships at local gyms, recreation centers, or health clubs. Others create or support a local recreation league, community sports, group team building, or recreational activities.

Offices should set aside places for uninterrupted thinking for contemplation of problems and to complete plans. It could be an outside space for walks or sitting undisturbed. A Stanford study calculated that taking a walk increased creativity by 60 percent.

Parkinson's Law is often used in reference to time usage. It states that the more time you've been given to do something, the more time it will take you to do it. It's amazing how much you can get done in 20 minutes if 20 minutes is all you have. But if you have all the time in the world—that's how long it may take to get things done.

SOURCES

"2014 Wasting Time at Work Survey." *Chron*, March 18, 2014, https://www.chron.com/jobs/salary/article/2014-Wasting-Time-at-Work-Survey-5347458.php.

"10 Hour Days vs. 8 Hour Days." Woodweb, August 13, 2002, http://www.woodweb.com/knowledge_base/10_hour_days_vs_8_hour_days.html.

"Americans Waste More Than 2 Hours a Day at Work, Costing Companies $759 Billion a Year, According to Salary.com and America Online Survey." *Business Wire*, July 11, 2005, https://www.businesswire.com/news/home/20050711005088/en/Americans-Waste-2-Hours-Day-Work-Costing.

Beard, Sienna. "Workers Are Wasting More and More Time on the Job." *Cheat Sheet*, October 27, 2014, https://www.cheatsheet.com/money-career/just-how-big-of-a-problem-has-wasting-time-at-work-become.html/?a=viewall.

Boyd, Drew. "Thinking Outside the Box: A Misguided Idea." *Psychology Today*, February 6, 2014, https://www.psychologytoday.com/us/blog/inside-the-box/201402/thinking-outside-the-box-misguided-idea.

Brennan, Ryan. "Half of Full-Time American Employees Spend More Time at Work Than with Family, Self." *Financial Juneteenth*, May 9, 2015, http://financialjuneteenth.com/half-of-full-time-american-employees-spend-more-time-at-work-than-with-family-self/.

Brown, Edward. "The Hidden Costs of Interruptions at Work." *Fast Company*, April 14, 2015, https://www.fastcompany.com/3044667/the-hidden-costs-of-interruptions-at-work.

"Calling in Sick: 7 Good Reasons, 7 Lame Reasons." Salary.com. https://www.salary.com/calling-in-sick-7-good-reasons-7-lame-reasons/.

Carmichael, Sarah Green. "The Research Is Clear: Long Hours Backfire for People and for Companies." *Harvard Business Review*, August 19, 2015, https://hbr.org/2015/08/the-research-is-clear-long-hours-backfire-for-people-and-for-companies.

Cho, Michael. "Why Do Some Employees Waste Time at Work?" Paychex, October 12, 2016, https://www.paychex.com/articles/human-resources/why-do-employees-waste-time-at-work.

Conner, Cheryl. "Employees Really Do Waste Time at Work." *Forbes*, July 17, 2012, https://www.forbes.com/sites/cherylsnappconner/2012/07/17/employees-really-do-waste-time-at-work/#7b2e3ffa5e6d.

Conner, Cheryl. "Wasting Time at Work: The Epidemic Continues." *Forbes*, July 31, 2015, https://www.forbes.com/sites/cherylsnappconner/2015/07/31/wasting-time-at-work-the-epidemic-continues/#702188961d94.

"Cyberveillance at Work." *CNN Money*, January 4, 2000, http://money.cnn.com/2000/01/04/technology/webspy/.

DeMers, Jayson. "Why the 40-Hour Workweek Is Dying." *Forbes*, May 15, 2015, https://www.forbes.com/sites/jaysondemers/2015/05/15/why-the-40-hour-workweek-is-dying/#2b33137649c6.

Deutschmann, Jennifer. "Employees Waste up to 80 Percent of Time 'Cyberloafing.'" *Inquisitr*, February 6, 2013, https://www.inquisitr.com/511795/employees-waste-up-to-80-percent-of-time-cyberloafing-study/.

"Do You Waste Time on Emails." American Society of Employers, September 27, 2017, https://www.aseonline.org/News/Articles/ArtMID/628/ArticleID/1275/Quick-Hits-September-27-2017.

Eadicicco, Lisa. "Americans Check Their Phones 8 Billion Times a Day." *Time*, September 15, 2015, http://time.com/4147614/smartphone-usage-us-2015/.

Emerson, Ramon. "Women Use Social Media More Than Men: Study." *Huffington Post*, December 26, 2017, https://www.huffingtonpost.com/2011/09/23/women-use-social-media-more_n_978498.html.

"Facebook & Too Many Meetings Top the List of Employee Time-Wasters." Salary.com, https://business.salary.com/why-how-your-employees-are-wasting-time-at-work/.

Fallon, Nicole. "What's Your Most Productive Work Time? How to Find Out." *Business News Daily*, July 11, 2017, https://www.businessnewsdaily.com/8331-most-productive-work-time.html.

Farber, Madeline. "Smartphones Are Making You Slack Off at Work." *Fortune*, June 6, 2016, http://fortune.com/2016/06/09/smartphones-making-you-slack-at-work-survey/.

Ferro, Shaunacy. "How Much Time Do We Actually Spend Working at Work? *Mental Floss*, February 1, 2016, http://mentalfloss.com/article/74710/how-much-time-do-we-actually-spend-working-workany.com/3044667/the-hidden-costs-of-interruptions-at-work.

Friedman, Eric. "10 Steps to Keeping Employees Engaged and Motivated." *eSkill* (blog), http://blog.eskill.com/employees-engaged-motivated/.

Green, Allison. "5 Things to Know about Taking Time off Work." *U.S. News and World Report*, July 17, 2013, https://money.usnews.com/money/blogs/outside-voices-careers/2013/07/17/5-things-to-know-about-taking-time-off-work.

Green, Harlan. "It's Time for the 30-Hour Week." *Huffington Post*, December 6, 2017, https://www.huffingtonpost.com/harlan-green/its-time-for-the-30-hour_b_7696674.htm.

Gould, Tom. "Workplace Follies: 11 Amazing Things Employees Did Instead of Their Jobs." *HRMorning*, July 25, 2014, http://www.hrmorning.com/workplace-follies-11-amazing-things-employees-did-instead-of-their-jobs/.

Gouveia, Aaron. "2014 Wasting Time at Work Survey." *Chron*, March 18, 2014, https://www.chron.com/jobs/salary/article/2014-Wasting-Time-at-Work-Survey-5347458.php.

"Here's Where Your Team Is Wasting Time at Work." *Time Doctor*, https://biz30.timedoctor.com/infograph-wasted-time/.

"How Bosses Say Workers Waste Time." *CNBC News*, https://dwc.cnbc.com/rki2r/?fs=1.

"How Much Does Pooping at Work Cost the US Economy Each Year?" Reddit, https://www.reddit.com/r/theydidthemath/comments/36ukrl/request_how_much_does_pooping_at_work_cost_the_us/.

Leadem, Rose. "Your Employees Are Wasting Time at Work, but Their Managers Slack Off Even More." *Entrepreneur*, August 10, 2016, https://www.entrepreneur.com/article/280638.

Lebowitz, Shana. "Here's How the 40-Hour Workweek Became the Standard in America." *Business Insider*, October 24, 2015, http://www.businessinsider.com/history-of-the-40-hour-workweek-2015-10.

Lebowitz,Shana. "The 40-Hour Workweek Is on Its Way Out." *Business Insider*, May 5, 2015, http://www.businessinsider.com/working-more-than-40-hours-a-week-2015-5.

Malenke, Kirsten. "Managing Employee Time-Wasting Activities." Advance Healthcare Network Jobs, http://advanceweb.com/jobs/healthcare-news/hr-employee-retention-articles/managing-employee-time-wasting-activities.htm.

Martin, Daniel, "3 Ways Millennials Waste Time without Knowing It." *Forbes*, August 7, 2017, https://www.forbes.com/forbes/welcome/?toURL=https://www.forbes.com/sites/under30network/2017/08/17/3-ways-millennials-waste-time-without-knowing-it.

Matthews, Steven. "Men Work Longer Hours Than Women: Male Employees Spend 40 Minutes More Time in the Office Than Female Colleagues." *Daily Mail*, June 27, 2016, http://www.dailymail.co.uk/sciencetech/article-3662333/Men-work-LONGER-hours-women-Male-employees-spend-40-minutes-time-office-female-colleagues.html.

MetLife Mature Market Institute. *The MetLife Study of Gen X: The MTV Generation Moves into Mid-Life*. April 2013, https://www.lifehappens.org/wp-content/uploads/2015/02/Research_MetLifeStudyofGenX-TheMTVGenerationMovesIntoMidLife.pdf.

Moore, Susie. "10 Reasons to Take a Random Day off Work." *Huffington Post*, June 9, 2014, https://www.huffingtonpost.com/susie-moore/day-off-work_b_5038387.html.

"New CareerBuilder Survey Reveals How Much Smartphones Are Sapping Productivity at Work." *Career Builder*, June 2, 2016, http://www.careerbuilder.com/share/aboutus/pressreleasesdetail.aspx?sd=6%2f9%2f2016&id=pr954&ed=12%2f31%2f2016.

"New Survey Shows Time's a Wastin'—Workers Goof Off More Than Two Hours a Day." American Management Association, http://www.amanet.org/training/articles/new-survey-shows-times-a-wastin-and-mdash;workers-goof-off-more-than-two-hours-a-day.asp.

"OECD Reveals Countries with Longest Working Hours," *Huffington Post*, April 24, 2012, https://www.huffingtonpost.com/2012/05/24/11-countries-with-the-longest-working-hours_n_1543145.html.

"One More Cringe-Worthy Stat about Small Business Websites." *Business Search Marketing*, April 17, 2012, http://www.smallbusinesssem.com/one-more-cringe-worthy-stat-about-small-business-websites/5674/.

O'Neill, Natalie. "Why We Should Quit the 40-Hour Work Week." *New York Post*, February 3, 2017, https://nypost.com/2017/02/03/why-we-should-quit-the-40-hour-work-week/.

Pappas, Stephanie. " How Big Is the Internet, Really?" *Live Science*, March 18, 2016, https://www.livescience.com/54094-how-big-is-the-internet.html.

"Parkinson's Law." Wikipedia, 2013, https://en.wikipedia.org/wiki/Parkinson%27s_law.

Patrick, Erin. "Employee Internet Management: Now an HR Issue." Society for Human Resource Management, https://shrm.org/hr-today/news/hrmagazine/Pages/cms_006514.aspx.

Peterson, Kim. "Your Phone Is Killing Your Work Productivity." *Money Watch*, June 12, 2014, https://www.cbsnews.com/news/your-phone-is-killing-your-work-productivity/.

Pozi, Ilya. "Are Employees Actually Doing Any Work?" *Inc.*, January 2, 2015, https://www.inc.com/ilya-pozin/are-employees-actually-doing-any-work.html.

"Report: workers spend 25% of work time goofing around online." *ars technical*, September 26, 2008, https://arstechnica.com/uncategorized/2008/09/report-workers-spend-25-of-work-time-goofing-around-online/.

"Sleep Deprivation." Wikipedia, https://en.wikipedia.org/wiki/Sleep_deprivation.

Treseder, William. "Social Media Cost Employers Billions in Lost Productivity." LinkedIn, September 15, 2014, https://www.linkedin.com/pulse/20140915221534-21875070-social-media-cost-employers-billions-in-lost-productivity.

Warner, Russ. "Who Wastes More Time at Work: Millennials, Gen X'ers or Boomers?" *Huffington Post*, April 7, 2013, https://www.huffingtonpost.com/russ-warner/who-wastes-more-time-at-w_b_2618279.htmlwork-become.html/?a=viewall.

"Web Surfing 'as Addictive as Coffee.'" *CNN News*, May 19, 2005, http://edition.cnn.com/2005/BUSINESS/05/19/web.work/index.html.

"Why We Should Kill the 40 Hour Work Week." *Crew*, December 21, 2017, https://crew.co/blog/why-you-shouldnt-work-set-hours/.

Wong, May. "Stanford Study Finds Walking Improves Creativity." *Stanford News*, April 24, 2014, https://news.stanford.edu/2014/04/24/walking-vs-sitting-042414/.

"Workplace Interruptions." 4imprint.com, http://info.4imprint.com/wp-content/uploads/1P-21-0915-Workplace-Interruptions.pdf?586da0.

18

PET WASTE—THE REAL POOP

The greatness of a nation and its moral progress can be judged by the way that its animals are treated.
—*Mahatma Gandhi*

It started in prehistoric times when wolves began hanging around humans for scraps. With farmers, it started with cats as a convenient rodent control. Over time, both discovered the value of companionship, and human, cat, and canine became best friends; far better living working beside them than having them jump at us out of the dark as predators.

Most people who have pets consider them members of the family. In fact, some people even prefer their pets to family members. And the affection of family pets, especially dogs, is limitless.

Oxytocin, sometimes referred to as "the brain's love drug" is increased as much as 40 to 60 percent when we see someone we love. In one study, researchers found that dogs, as well as humans, produce large amounts of oxytocin while playing and interacting, or just looking into each other's eyes.

Researchers at the University of Oregon offered 38 cats a choice between food, a toy, a tempting smell, or playtime with a human. More than half the cats chose the human. Dogs, as we all know, have the ability to express unconditional love.

Pets, especially dogs, rescue humans all the time—there are even medals awarded to animals that save people. Unfortunately, due to our neglect, millions of animals die every year due to euthanasia, medical research practices, bad breeding habits, and homelessness.

HORRIBLE HISTORY

In some ways, the history of animal shelters in the United States is embarrassing. Early animal control was little more than a hostage-taking operation, and excess animals were considered waste, or simply a resource to be exploited. Pounds held wandering animals for ransom from their owners. Unclaimed cows and pigs went to the highest bidder. Because dogs had little economic value, the pound-master typically executed them using whatever means he found convenient and cheap. New York City developed a grisly solution to its stray dog problem in the 1870s: It locked dozens of dogs at a time inside an enormous iron cage and submerged them in the East River—capture, drown, repeat—until the population was under control. No one objected to this final solution.

Until recently, almost all shelters would kill the old to make room for newer and more adoptable animals. And many of the municipal shelters, which are obliged by law to accept all animals, continue to euthanize 50 percent or more of the animals they take in.

Today, 70 percent of American pet owners believe unwanted dogs, unless they are incurably ill or irredeemably aggressive should be cared for indefinitely.

Repugnant Postmortem Profits?

All the wholesome nutrition your dog or cat will ever need comes in your pet's foods. Right? Well, that's the image the $24 billion U.S. pet food industry wants pet owners to believe. What the boutiquey ads don't tell you is that pet food can mean anything from grains considered "unfit" for human consumption to (according to the National Renderers Association) offal, intestines, udders, heads, hooves, and other parts of a recently alive mammal from the stockyards, road kill, or even somebody's recently deceased pet. FYI, some of those rendered products also go into your cosmetics, soap, and even toothpaste.

Urban legends have circulated a persistent rumor that some pet foods are made, in part, from the dead remains of pets that can be profitably put to use as pet fodder. There are disturbing bits of evidence that may prove that it was done in the past and is still the case in the present.

If you're into checking the labels on grocery store foods, you're hip to the fact that the list of ingredients doesn't always tell the whole truth about what's in your food and the same applies to pet food. Terms like "meat by-products" and "meat meal" might mean that what had once been a pet cuddling up for affection is now being offered as food for Fluffy or Fido. So-called 4D animals—dead, dying, diseased, disabled—are still legit ingredients for pet food.

When a vet tells a grieving "pet parent" that they'll "take care" of their dead loved one, they usually mean sending it off with the disposal company for rendering. Although many in the pet food industry deny that they use euthanized animals, which is perfectly legal, the fact is veterinarians and especially shelters usually don't have the money to bury or cremate animals.

Pet food executives have denied the charges. Nevertheless, state agencies, the FDA, and medical groups such as the American Veterinary Medical Association (AVMA) and the California Veterinary Medical Association confirm that pets, on a routine basis, have been rendered after they die in animal shelters or are disposed of by health authorities, and the end product may frequently find its way into pet food.

California, for example, allows rendered pets to be processed and sold out of state for pet food as meat and bone meal. The city of Los Angeles alone sends about 200 tons of dead pets to a rendering plant each month. By the way, animal fat is that pungent smell you notice when you open a new bag of dry dog or cat food; the kibble is sprayed with it to make it more palatable to your pet.

A story in the *Los Angeles Times* in 2002 said, "By far the bulk of rendered material comes from slaughterhouses, the trimmings from supermarkets, road kill, dead farm animals, and euthanized pets from shelters."

It's a nice, neat, profitable daisy chain. Shelters send their euthanized animals to renderers, who turn the animals into "ingredients," which are then sold to the pet food companies and shipped to the shelters to feed live animals. It's kind of like Soylent Green for pets.

Facts and Figures:

- Sixty-eight percent of U.S. households, or about 85 million families, own a total 90 million cats and dogs, according to the current National Pet Owners Survey conducted by the American Pet Products Association (APPA).
- Approximately eight million companion animals enter U.S. animal shelters nationwide every year. Of those, approximately 3.3 million are dogs and 3.2 million are cats. More than half won't make it out alive. Approximately 3.2 million shelter animals are adopted each year (1.6 million dogs and 1.6 million cats).
- Over $2 billion is spent annually by local governments to shelter and ultimately destroy adoptable dogs and cats due to shortage of homes.
- On average, it costs approximately $100, much of which comes out of our pockets, to capture, house, feed, and eventually kill a homeless animal, states the Doris Day Animal League.

Spay and Neuter

The cost of having a pregnant female give birth is much higher than the cost of spaying. There's absolutely no truth to the myth that it's best to let a female pet give birth to a litter before getting her spayed. Just one female dog and her puppies can result in 67,000 dogs in six years, and one female cat and her kittens can lead to 370,000 cats.

Spay U.S. points out that each day 10,000 humans are born in the United States along with 70,000 puppies and kittens. Do the math.

THE PET TEST

One of the prickliest questions for anyone who cares about animals is whether one species can use another for testing. Is it worth animal lives to find a new and better mascara? How about cure for a deadly disease? Most people think the first question frivolous, but the second important.

The term *animal testing* refers to procedures performed on living animals for purposes of research into biology and diseases, assessing the effectiveness of new medicinal products, or the environmental safety of consumer products from pharmaceuticals to cleaners and cosmetics. The USDA's 2015 annual report on animal use at research facilities shows a continued decreasing trend in the number of animals used in U.S. laboratories.

The report revealed that 904,147 animals covered by the Animal Welfare Act (AWA) were held in labs last year, and that 767,622 were used in research, a drop of over 8 percent from 2014. Hamsters are among the most used animals in labs, and in experiments involving pain. The "all other covered species" animal category, which includes gerbils, bats, ferrets, and chinchillas, decreased 24 percent. Also down were the 8 percent of rabbits and 5 percent of cats.

Unfortunately, there were increases in the number of other animals used in experiments. Primate numbers increased 7 percent, and there was a slight rise in the number of dogs, guinea pigs, and "other farm animals" (sheep, pigs, goats, cows, horses), used in research.

A Pew Research Center poll found that 50 percent of U.S. adults oppose the use of animals in scientific research.

Animal Testing—Pros and Cons

Pros

- Animal testing has contributed to many life-saving cures and treatments. The California Biomedical Research Association states that nearly every medical breakthrough in the last 100 years has resulted directly from research using animals.
- There is no adequate alternative to testing on a living, whole-body system.
- Animals are appropriate research subjects because they are similar to human beings in many ways. Chimpanzees share 99 percent of their DNA with humans, and mice are 98 percent genetically similar to humans.
- Animals must be used in cases when ethical considerations prevent the use of human subjects.
- Animals themselves benefit from the results of animal testing.

- The American Veterinary Medical Association (AVMA) endorses animal testing.
- Animal research is highly regulated, with laws in place to protect animals from mistreatment, and the Association for Assessment and Accreditation of Laboratory Animal Care International (AAALAC) reviews researchers for humane practices.
- Animals often make better research subjects than human beings because of their shorter life cycles.
- Animal researchers treat animals humanely, both for the animals' sake and to ensure reliable test results. Research animals are cared for by veterinarians, husbandry specialists, and animal health technicians to ensure their well-being and more accurate findings.
- The vast majority of biologists and several of the largest biomedical and health organizations in America endorse animal testing. A Pew Research Center found that 89 percent favored the use of animals in scientific research.
- Some cosmetics and health care products must be tested on animals to ensure human safety.
- Relatively few animals are used in research, which is a small price to pay for advancing medical progress.
- Animals do not have rights; therefore it is acceptable to experiment on them.
- Religious traditions allow for human dominion over animals.

Cons

- Animal testing is cruel and inhumane according to Humane Society International.
- Alternative testing methods now exist that can replace the need for animals.
- Animals are very different from human beings and therefore make poor test subjects.
- Drugs that pass animal tests are not necessarily safe. The 1950s sleeping pill thalidomide, which caused 10,000 babies to be born with severe deformities, was tested on animals prior to its commercial release.

- 95 percent of animals used in experiments are not protected by the Animal Welfare Act.
- Animal tests do not reliably predict results in human beings. Ninety-four percent of drugs that pass animal tests fail in human clinical trials.
- Animal tests are more expensive than alternative methods and are a waste of government research dollars.
- Most experiments involving animals are flawed, wasting the lives of the animal subjects.
- Animals can suffer like humans do, so it is "species-ism" to experiment on them while we refrain from experimenting on humans.
- The Animal Welfare Act has not succeeded in preventing horrific cases of animal abuse in research laboratories.
- Medical breakthroughs involving animal research could still have been made without the use of animals.
- Religious traditions tell us to be merciful to animals, so we should not cause them suffering by experimenting on them.
- Many believe that animals are entitled to the possession of their own lives and that the right to avoid suffering should be afforded the same consideration as human beings.

THE HOMELESS

More than 70 million stray animals live in the United States. Of the animals who wind up in a shelter, only half will be adopted into a home. The sad reality is that the other three or four million animals in shelters will be euthanized due to lack of space and funds. Meanwhile, many will live in wretched circumstances, reproducing recklessly, leading to further chaos.

Although the rates have continued to decline in recent decades, more than 2.7 million abandoned animals are still killed in this country each year. A recent study by Best Friends Animal Society found that half of Americans surveyed believed that approximately 500 dogs and cats are euthanized every day in shelters across the country. In reality, the number is almost ten times that much—a staggering 9,000 dogs and cats are killed at shelters daily due to the fact that they don't have a home and the shelters don't have the space.

Fifty-six percent of dogs and 71 percent of cats that enter animal shelters are euthanized, according to the American Humane Society. More cats are euthanized than dogs because they are more likely to enter a shelter without any owner identification. Only 25 percent of dogs and 24 percent of cats that enter animal shelters are adopted.

Rich Avanzino, president of the California-based Maddie's Fund, pioneered no-kill in San Francisco in the early 1990s through a pact between the city shelter and the local Society for the Prevention of Cruelty to Animals (SPCA). He is one of the advocates that points out that in this country tax payers spend $1 billion dollars annually to pick up, house, and euthanize homeless animals. As of 2015, Charity Navigator listed 90 major U.S. animal shelters with annual budgets over $3.5 million—all of these shelters together spent $1.2 billion per year in 2015.

Overpop Fallacy

"Over-population is a myth," says attorney Nathan Winograd, whose recent book, *Redemption: The Myth of Pet Overpopulation and the No-Kill Revolution in America* chronicles the rise of the no-kill shelter movement, and states, "With better outreach and public relations, we can find homes for virtually all of the healthy animals we are now killing."

Bonney Brown, executive director of the Nevada Humane Society, says that in 2007, the first year her group went "no-kill," her shelters managed to save 90 percent of the 8,000 animals they took in.

THE PETA USE/MISUSE OF MILLIONS

In 2014–2015, People for the Ethical Treatment of Animals (PETA) took in $51,933,001, in contributions; $627,336 in merchandise sales; and $856,642 in interest and dividends. They finished the year with $4,551,786 more "bank" than they started with. They did not see fit to use a lot of that to comprehensively promote animals for adoption or to provide veterinary care for the animals that needed it. But they did manage to do away with 72 percent of the animals they "collected," calling it "mercy killing."

By contrast, the Humane Society takes in more animals than PETA (they do not operate traditional shelters) and managed to save 94 percent—without PETA's mega-millions.

PETA's Projects:

- Grants to convicted arsonist Rodney Coronado and the eco-terrorist group Earth Liberation Front. Coronado and former PETA campaign coordinator Bruce Friedrich told an animal rights convention that "blowing stuff up and smashing windows is a great way to bring about animal liberation," adding, "Hallelujah to the people who are willing to do it."
- A "Vegans Make Better Lovers" campaign where PETA campaigners publicly made out on a bed on the streets of Nashville, Tennessee.
- A campaign comparing pregnant women to fattened sows to protest farmed meat.
- PETA's anti–farmed meat stunt featured a naked women in a cage.
- A campaign titled: "Holocaust on Your Plate," comparing animal use to the slaughter of six million Jews during World War II.
- The purchase of a cemetery plot located near the deceased Colonel Sanders with a headstone that reads KFC Tortures Birds.
- A campaign to urge Ben and Jerry's ice cream to drop cow milk in favor of human milk.

PETA vs. Pro-Life

Critics say that surely PETA can do better than their one-percent adoption rate, especially given their national media reach and enormous resources. And lately the protester is being protested, as PETA has become the number one target among supporters of no-kill shelters.

In contrast, Seagoville Animal Services in Texas took in about 700 animals, with only a small fraction of the amount of money that PETA received, and saved 99 percent of the animals on a paltry $29,700 budget. In fact, hundreds of cities and towns across America are saving over 90 percent of the animals they take in.

More than any other group, Maddie's Fund, a foundation in the San Francisco Bay Area with a $300 million endowment, has been responsible for spreading the no-kill movement. It was started in 1999 by Cheryl and David A. Duffield, dot-com billionaires, and named after their miniature schnauzer. They have financed medical training programs for shelters at 18 of the country's 29 veterinary schools, the idea being that healthy animals are cheaper to house and are more likely to be adopted.

Each year, the foundation sponsors adoption weekends in several cities. As an incentive, Maddie's Fund paid the shelters $500 for each dog and cat under seven years that was adopted, $1,000 for each animal over seven, and $2,000 for each animal over seven with a medical ailment. It spent $3 million to subsidize the 3,104 adoptions.

This doesn't impress PETA. Daphna Nachminovitch, the group's senior vice president of cruelty investigations, has said, "Adoption can be bad—far worse than euthanasia. We can offer them a painless release from a world that doesn't have enough room for them in its heart or homes." And their actions spoke as loud as their words, euthanizing the overwhelming majority of dogs and cats that it accepted into its "care."

PETA even claims that "euthanasia is a product of love for animals who have no one to love them" and calls their killing a "tragic reality" and one that acknowledges how "sometimes animals need the comfort of being put out of their misery—a painless release from a world in which they were abused and unwanted."

In 2007, two PETA employees were tried for animal cruelty and littering in North Carolina after they were caught in a late-night stake-out dumping the bodies of dead dogs and cats in a dumpster. Evidence presented during the trial showed that PETA employees killed animals they considered "adorable" and "perfect."

What's worse was PETA members' kidnapping of a perfectly healthy and young Chihuahua, Maya, from the Zarate family in Virginia. When Mr. Zarate's niece contacted PETA about the responsibility for Maya's disappearance the next day, PETA denied even being in the area at the time. When the owner told the representative at PETA that she had security footage of the dognapping by PETA people, the representative abruptly hung up.

Two days later, the PETA employees who had abducted Maya returned with a fruit basket and the news that Maya was dead. The family has since settled a lawsuit against PETA, who states, that it "considers pet ownership to be a form of involuntary bondage," much like slavery. Apparently, they think the animals are better off dead than adopted.

PETA's quasi-religious campaigns include a website that claims—despite ample evidence to the contrary—that Jesus Christ was a vegetarian. PETA's billboards taunt Christians with the message that hogs "died for their sins."

PETA President Ingrid Newkirk told *Vogue* that, "Even if animal tests produced a cure for AIDS, we'd be against it."

SOME SOLUTIONS

The American Humane Society believes the percentage of animals reunited with their owners would greatly increase if more pets were properly identified. So:

- Be sure your pet wears an identification tag, rabies license, and city license. Include your name, address, phone number, and pet's name.
- Keep licenses current, as they help shelters locate pet owners.
- When moving, or during a natural disaster, put a temporary tag on your pet including a phone number of someone who will know how to reach you.
- Don't assume that your indoor pet doesn't need tags—pets can escape.
- Purchase special cat collars with elastic bands to protect your cat from being caught in trees or on fences.
- In addition to ID tags, consider getting your pet microchipped, tiny electronic devices that are injected under the skin with a hypodermic needle. When a scanner is passed over a pet's skin, it emits radiofrequencies that activate the chip, which transmits a unique identification number that locates the owner.
- In a study of animal shelters, only 22 percent of dogs without microchips were reunited with their owners, while 52 percent with microchips returned home. Cats without microchips had even lower return rates: just 2 percent made it home, compared with 39 percent of microchipped felines.

- During natural disasters, pay close attention to your pets so they don't freak out and run away.

Our attitudes toward our animals reflects our attitudes toward life itself. Perhaps one of our most revered presidents, Abraham Lincoln, said it best, "I care not for a man's religion whose dog and cat are not the better for it."

SOURCES

"7 Things You Didn't Know about PETA." *Snugglenugget*, December 4, 2008, https://snugglenugget.wordpress.com/2008/12/04/7-things-you-didnt-know-about-peta/.

"8 Things You Didn't Know about PETA." PETA Kills Animals, https://www.petakillsanimals.com/wp-content/uploads/2016/03/PKA_Shareable.pdf.

"16 Integral Pros and Cons of Animal Experimentation." *Connect US*, https://connectusfund.org/16-integral-pros-and-cons-of-animal-experimentation.

"10 Pros and Cons of Animal Experimentation." *Flow Psychology*, https://flowpsychology.com/10-pros-and-cons-of-animal-experimentation/.

"23 Curious Facts about Dogs and Cats." *Down Home Pets*, http://www.downhomepets.com/domestic-pets/dogs/23-curious-facts-about-dogs-and-cats/.

"27 Famous Abraham Lincoln Quotes." *Bright Drops*, http://brightdrops.com/abraham-lincoln-quotes.

"All about Animal Testing, *SPCA*, http://www.hsi.org/campaigns/end_animal_testing/qa/about.html.

"Which Animals Are Used." *American Anti-Vivisection Society*, http://aavs.org/animals-science/animals-used/.

"All the Animal Testing Pros and Cons That You Never Knew About." *Opinion Front*,

"Animal Shelter Euthanasia." *American Humane*, https://www.americanhumane.org/fact-sheet/animal-shelter-euthanasia-2/.

"Ethics of Medical Research with Animals." Hastings Center, http://animalresearch.thehastingscenter.org/facts-sheets/animals-used-in-research-in-the-united-states/.

Cathcart, Nancy. "12 Vermont Shelters and Rescue Groups Join Maddie's Pet Adoption Days." Vermont Chamber of Commerce, April 6, 2014, https://vtdigger.org/2014/04/06/12-vermont-shelters-rescue-groups-join-maddies-pet-adoption-days/.

"Does Pet Food Include Rendered Euthanized Pets?" Snopes, https://www.snopes.com/fact-check/pets-in-pet-food/.

Douglass, Anne. " Does PETA Have the Right to Determine What's "Humane" Considering Their View on Animals?" Certified Humane, January 13, 2016, https://certifiedhumane.org/does-peta-have-the-right-to-determine-whats-humane-considering-their-view-on-animals/.

"Facts and Figures." SPCA, https://media.rspca.org.uk/media/facts.

Greenwood, Arin. "PETA Euthanized a Lot of Animals at Its Shelter in 2014, and No-Kill Advocates Are Not Happy about It." *Huffington Post*, December 6, 2017, https://www.huffingtonpost.com/2015/02/05/pets-shelter-euthanization-rate_n_6612490.html.

Hanna-Funk, Jasmine. "Why PETA Sucks." *Prezi*, April 19, 2015, https://prezi.com/qs4aptm5lyjv/why-peta-sucks/.

Harpst, Lynne Koen. "Who Rescued Whom." *Coronado Clarion*, October 3, 2012, http://www.coronadoclarion.com/winter-2013/.

Jenkins, Beverly. "10 Disgusting Common Ingredients in Cosmetics." Oddee, September 14, 2012, https://www.oddee.com/item_98322.aspx.

Interlandi, Jeneen. "PETA and Euthanasia." *Newsweek*, April 27, 2008, http://www.newsweek.com/peta-and-euthanasia-85753.

Lee, Rainie, and Carrie Funk. "Opinion about the Use of Animals in Research" The Pew Research Center, July 1, 2015, http://www.pewinternet.org/2015/07/01/chapter-7-opinion-about-the-use-of-animals-in-research/.

Lehnardt, Karin, "100 Interesting Facts about Cats." Fact Retriever, August 19, 2016, http://www.downhomepets.com/domestic-pets/dogs/23-curious-facts-about-dogs-and-cats/.

"Pit Bull Bans: The State of Breed Specific Legislation." Animal Legal Defense Fund, July 31, 2009, http://aldf.org/press-room/press-releases/pit-bull-bans-the-state-of-breed-specific-legislation/.

McWilliams, James. "PETA's Terrible, Horrible, No Good, Very Bad History of Killing Animals." *Atlantic*, March 12, 2012, https://www.theatlantic.com/health/archive/2012/03/petas-terrible-horrible-no-good-very-bad-history-of-killing-animals/254130/.

Newkirk, Ingrid. "Why We Euthanize." PETA, February 2018. https://www.peta.org/blog/euthanize/.

"No Kill Advocacy Center." Wikipedia, https://en.wikipedia.org/wiki/No_Kill_Advocacy_Center.

"No-Kill-Shelter Activists Target Holdout PETA." *Seattle Times*, July 6, 2013, https://www.seattletimes.com/life/pets/no-kill-shelter-activists-target-holdout-peta/.

O'Hare, Tim. "Factors to Consider before You Compare Pet Insurance?" Pet Place, April 11, 2018, https://www.petplace.com/article/general/pet-insurance/factors-consider-compare-pet-insurance-policies/.

"Overpopulation in Domestic Pets." *InfoGalactic*, https://infogalactic.com/info/Overpopulation_in_domestic_pets.

"Over population of Animals." Revolvy, https://www.revolvy.com/topic/Overpopulation%20animals&item_type=topic.

Palmer, Brian. "Are No-Kill Shelters Good for Cats and Dogs?" *Daily Independent*, August 5, 2014, http://www.dailyindependent.com/are-no-kill-shelters-good-for-cats-and-dogs/article_b88d3678-3fd5-5903-bfbf-dacc2ccf04af.html.

"Pet Euthanasia in Shelters Unpopular." *CBS News*, January 5, 2012, https://www.cbsnews.com/news/pet-euthanasia-in-shelters-unpopular/.

"Proof PETA kills." *PETA Kills Animals*, https://www.petakillsanimals.com/proof-peta-kills/.

"Pros and Cons of Animal Testing." Health Research Funding, February 26, 2014, https://healthresearchfunding.org/pros-cons-animal-testing/.

"Should Animals Be Used for Scientific or Commercial Testing?" Pro and Con, https://animal-testing.procon.org/.

"Size of the U.S. Pet Food Market in 2016 and 2022." Statista, https://www.statista.com/statistics/755068/us-pet-food-market-size/.

Smith, Jessica. "The True Horrors of Pet Food Revealed: Prepare to Be Shocked by What Goes into Dog Food and Cat Food." *Natural News*, October 21, 2005, https://www.naturalnews.com/012647.html.

Stanley, Coren. "How Dogs Were Created." *Modern Dog*, https://moderndogmagazine.com/articles/how-dogs-were-created/12679.

Stratton, Lynn."What Do You Really Know about Rendered Pet Food?" *Healthy Holistic Living*, http://www.healthy-holistic-living.com/rendered-pet-food.html.

"The Dog Rescuers." https://www.thedogrescuers.com/statistics--facts.html.

"The Unshakeable Love between Pets and Owners." Elmhurst Animal Care Center, January 24, 2018, http://www.elmhurstanimalcarecenter.com/blog/pets-and-owners/#.WupIKDNlDcs.

Team Darwin. "The Hidden Dimension in Pet Food: 3D and 4D Meats." Darwin's Natural Pet Products, July 20, 2016, https://www.darwinspet.com/ask-darwins-3d-and-4d-meats/.

"USDA/APHIS Finalizes Rule Impacting Pet Breeders." American Kennel Club, Northern California Working Group Assoc, Inc., from *The New York Times*, September 10, 2013, http://www.ncwga-inc.org/CA%20Dog%20legislation2.html.

Torchia, Michael. "The Sickening Truth about Pet Food." *News Blaze*, March 1, 2008, https://newsblaze.com/issues/animals/the-sickening-truth-about-pet-food_4110/.

"Very Disturbing News Concerning the People for the Ethical Treatment of Animals PETA." *Catholic Online*, February 28, 2012, https://www.catholic.org/news/politics/story.php?id=44940.

"PETA says Sorry for Taking Girl's Pet Chihuahua and Putting It Down." *Guardian*, https://www.theguardian.com/us-news/2017/aug/17/peta-sorry-for-taking-girls-dog-putting-it-down.

"Why Farmed Animals?" Animal Charity Evaluators, November 2016, https://animalcharityevaluators.org/donation-advice/why-farmed-animals/.

"Why No-Kill Animal Shelters Are Good. No, Bad. No, Good." *Tampa Bay Times*, May 19, 2014, http://www.tampabay.com/news/nation/why-no-kill-shelters-are-good-no-bad-no-good/2180441.

19

TREASURE FROM TRASH

Appreciate what you have before it is what you had.

Most of us have been tantalized or traumatized by what we have thrown away, or what someone else has tossed out, something we passed over, missed out on, or absentmindedly let go—and found out too late that it was treasure. We're reminded by TV shows like *Antiques Roadshow* or *American Pickers*, which stimulate our fantasies about something in the attic, garage, or yard sale down the street that could be worth a small or even large fortune.

It makes us think back about valued relics of another time, whether it was vinyl records, Hummel figurines, baseball cards, Beanie babies, Barbie dolls, or any one of millions of products that are now collector's items. Boomers and even Millennials are collecting remembrances of things past—although some of these items are getting rather pricey. The worldwide value of the collectible toy markets is said to be $10 billion per annum, according to toy experts from the hobbyDB project.

The old truism that one man's trash is another man's treasure is spot on in many ways. A primo example is a museum on the second floor of an East Harlem, New York, sanitation garage. It's one of the most eclectic collections in the world—a gallery where you might say that all the contents are garbage—and you'd be absolutely right.

It's the culmination of 30 years of picking through New York City rubbish by one of its own, Nelson Molina, a 30-year vet of the NYC sanitation department, who nicknamed his compilation, "The Treasures in the Trash Museum."

Molina started his collection from his office locker room in 1981 because although he was finding stuff that interested him on his rounds, sanitation workers aren't allowed to bring items home. Soon, other sanitation workers on his route began to contribute pieces. As his fame grew, workers across the city started to bring him interesting objects rescued from the refuse, and building superintendents would save odd items for him. Today, Molina estimates his collection is 1,000 pieces large and still growing. "There are amazing things that people throw out, that people just don't want."

The Treasures of the Trash Museum contains just about anything from Furbies to old family photos to vintage war uniforms, typewriters, toys, ceramics, chairs, cameras, memorabilia, vacuum cleaners, and even a Star of David forged from the metal of the World Trade Center rubble commemorating a man who lost his life on 9/11.

The collection was even examined by the head curator of television's *Antiques Roadshow*, who proclaimed it one of the most diverse mixes she had ever seen.

"One man's trash is another man's treasure" is Molina's mantra, and he adds, "it doesn't matter what it is. As long as it's cool, I can hang it up and I've got a place for it."

The trash museum isn't open to the public, but you can schedule a visit by emailing the NYC Department of Sanitation at tours@dsny.nyc.gov.

REDUCING REFUSE BY REUSING

When making something new out of something old only the imagination is the limit. Clever people make the most incredible things out of what most of us just toss out. Old CD racks into planters, buttons made into bowls, egg cartons that store cupcakes, broken plate mosaics, birdfeeders from empty bottles, there's literally no end to what you can do with trash and your creative juices, and perhaps making a little cash while you're at it.

Tom Szaky, CEO of TerraCycle, is the ultimate eco-capitalism junky. "I don't see garbage anymore. I just see cash," he said. "This is an eco-revolution and we are eco-capitalists where we can make a boatload of money and save the world at the same time." He does this by upcycling, which is the art of turning garbage into making not only something new, but money, too.

He has managed to capitalize by turning used cookie wrappers into kites and first got inspired by selling worm feces for fertilizer when he was in college at Princeton. After he dropped out, he started his company, which now has three branches—in New Jersey, Toronto, and Atlanta.

And corporate America has taken notice. Mars, the world's leading manufacturer of chocolate , chewing gum, mints, and fruity confections, has partnered with TerraCycle to produce affordable, high-quality consumer goods by repurposing surplus and used packaging from more than 20 Mars brands. He's done well enough that he is able to offer people money for their garbage.

TALES OF TRASH INTO CASH

Along with winning the lottery are dustbin daydreams of finding something that brings a big hit with a little effort. And every once in a while it happens.

People who were getting ready for an estate sale came across an old and foreign-looking car. Out of curiosity, they called someone to identify the make and model. It was a rare 1937 Bugatti Type 57S Atalante with a mere 24,000 miles and all original parts. Only 17 were ever made. At auction, it went for around $3 million.

A Pennsylvania man purchased a painting for $4 at a local flea market because he liked the frame and thought he could restore it. He took it apart, but decided that it was too far gone. While throwing out the frame he discovered an old folded up piece of paper behind the picture. It turned out to be one of 24 known copies of the Declaration of Independence. It sold at auction for $2.42 million.

A New York woman stumbled upon a painting on a Manhattan street waiting for the garbage collectors. She liked it and took it home. The woman spent four years trying to find out about the painting. She finally discovered the artist on the *Antiques Roadshow* website. It turned out to be a stolen masterpiece called *Tres Personajes* painted by the famous Mexican artist, Rufino Tamayo. She received a $15,000 reward for turning in the painting and also received a percentage of the sale price of $1,049,000.

A Massachusetts man that regularly dove into convenience store dumpsters looking for lottery tickets thought he hit the jackpot when he found a $1 million winner. But the man who originally purchased the winning ticket later sued him, and he eventually had to give up $140,000 of his winnings in a settlement.

A reverend at a Catholic university was looking for paper towels in a bathroom cabinet and pulled out a frame containing an etching. It was identified as a Rembrandt worth about $100,000.

A British farmer lost a hammer in his fields one day and borrowed a metal detector to find it. Instead, he found a cache of Roman Empire–era artifacts containing 15,000 coins of various metals, including gold, plus jewelry and statues. He sold them for £1.75 million, which he generously split with the man who lent him the metal detector.

A man stopped at a yard sale on his way home one day and bought a painting, which he used to cover a hole in the wall of his home. A few years later, he was playing a board game with an art theme called "Masterpiece" and noticed that one of the paintings looked like his hole-in-the-wall painting. After some research, he found that the art was an original by Martin Johnson Heade, a classical American still-life artist. It was bought by a museum for $1.2 million.

Abandoned storage units are auctioned off to the public all the time and most contain a previous owner's junk. A San Jose man paid $1,100 for a unit that contained $500,000 worth of gold bars, silver, and rare coins.

A 72-year-old California woman was cleaning out a section of her home when she came across an old baseball card for which she was going to ask $10. A friend talked her into having an expert look at the card. He identified it as from 1869, and it was in near-perfect condition. It auctioned for more than $75,000.

A Kentucky man who worked at a recycling center found $22,000 worth of U.S. savings bonds lying in a barrel of scrap metal. More surprisingly, he tracked down the rightful owner of the savings bonds and returned them to him.

A handyman was hired to clean out the apartment of an artist who had recently passed away. He discovered a cardboard barrel that he decided to keep and put it away in a warehouse where it stayed for the next few years. Out of curiosity he opened the barrel to see what was inside. To his surprise, he found ancient Mayan artifacts dated between 300 B.C. and 500 A.D., worth $16,500.

One garbage man in Las Vegas found enough casino chips and other valuables that he was able to buy a home. What happens in Vegas really stays in Vegas.

One of Steve Jobs's first-ever computers, the 1976 Apple 1, ended up at a recycling center in California. They sold it at auction for $200,000, and the manager of the recycling center has been trying to track down the woman who brought it in so that she can claim her half of the fortune—very nice of them indeed.

A reporter that was asked to come on a treasure hunt in the Sierra foothills helped uncover the Goat Herder's cache—thousands of dollars' worth of paper bills and silver coins dating back decades.

One couple inherited two paintings that they loathed and kept in their basement for more than a decade. They were about to toss them when they decided to see whether they were worth anything. For one of the paintings of an Arizona desert they were happy to get $1,100. The art dealer who looked at their paintings asked if they would sell the other abstract "monstrosity" for more than $300,000. Luckily, they decided to wait and took it to auction. It was a lost masterpiece and brought $650,000.

And then there were the sanitation workers who heard some screams coming from the back of their truck—it was a guy that had fallen asleep back there and was about to be squashed.

So the cautionary tale here is to watch what you throw out, and check out what might be for sale around the corner. Few of us are going to make a whopping big strike, but there is a good chance that we can pick up more than spare change just by watching, waiting, and paying attention for potential treasures in the trash. The internet is a great place to start, whether you're looking to collect or to sell whatever your head or heart desires.

NOT SO GOOD NEWS

Then there is the story about the first journalist to ever interview J. K. Rowling. He received a first edition copy of *Harry Potter and the Phi-*

losopher's Stone as a thank-you for a review. Convinced it would flop, the reporter threw it away. Hopefully someone found it—first edition copies sell for up to £50,000 (about $67,500).

Diamonds might be forever. Just ask the person who accidentally threw $5 million worth of them in the trash. A "lucky" security guard found the treasure, but his luck quickly ran out when he was caught trying to sell them on the black market.

A husband gave his wife a surprise she'll never forget for their 23rd anniversary—he cleaned out her closet. However, he also accidentally threw away $50,000 worth of her jewelry. That's an anniversary they'll never forget.

There are legions of legends about people plucking plunder from trash. A group of workers cleaning a Washington State garbage dump discovered a glass bottle full of plutonium—the same plutonium used to make atomic bombs. They had to turn it in, but were lucky that it didn't harm anyone, and weren't arrested.

A discarded laptop found lying amid trash in a dumpster contained secret emails and other information related to the controversial sale of mortgage securities. The information helped convict a trader of mortgage fraud.

What Collectibles to Watch Out For

We're not talking prized antiques, vintage jewelry, or collectors cars, but just stuff you might run across at a thrift store or find in the attic. Basically anything collectable is any object regarded as being of value or interest to a collector.

As with all articles, condition is the determination of worth along with other considerations like being in an unopened or original box, whether a collection is complete, and what it will mean to you and others.

Online marketplaces like eBay, Etsy, even craigslist offer channels for buyers and sellers. Old favorites to haunt and hunt are antique shows, flea markets, yard and estate sales, and church bazaars that present opportunities to buy at medium to rock-bottom prices.

Scrap Prices Soar

Biodiesel, an environmentally friendly alternative made from cooking oil, has been commanding record prices. However, prices for the stuff have risen so much in recent years that thieves are now "casing" fast-food restaurants to plot grease heists, surreptitiously draining the vats of fat behind fast-food joints.

Scrap metal prices have also skyrocketed, especially copper, so much so that thieves are tearing apart street lights for the wire. The price of scrap metal trades just like a stock market does in the United States. This means the spot price that you might receive for your scrap metals today could be very different tomorrow.

DON'T LET THE THINGS YOU OWN END UP OWNING YOU

There is a huge gap between being lazy, offering excuses for not throwing away excess stuff, and being obsessive about letting things go. You gotta be tough, you gotta be merciless, you gotta be determined—to rid yourself of clutter. There's a great feeling of liberation that can come from shedding superfluous items. Grab a big bin and some bags and start putting things in order to toss, donate, give away, or recycle. Someone might be looking for something you want to throw away. But check for valuables first.

SOURCES

Anders, Drew. "8 Insanely Valuable Items You Probably Owned and Threw Out." Cracked, September 24, 2014, http://www.cracked.com/article_21498_8-insanely-valuable-items-you-probably-owned-and-threw-out.html.

Bazarraa, Danya. "Rare 1976 Apple Computer Worth $200,000 Accidentally Thrown Away at Recycling Centre." *Mirror*, May 13, 2015, https://www.mirror.co.uk/news/world-news/rare-1976-apple-computer-worth-5797471.

Benzaemon. "A List of Really Valuable Stuff You Might Actually Have in Your Loft." *Sabotage Times*, April 15, 2016, https://sabotagetimes.com/life/a-list-of-really-valuable-stuff-you-might-actually-have-in-your-loft.

Crews, Barb. "13 Ways to Sell Your Stuff." ThoughtCo, June 24, 2018, https://www.thoughtco.com/how-to-sell-your-stuff-782021.

Duncan, Lindsay. "Ouch! People Have Lost Thousands by Throwing Out These Valuable Items." June 23, 2016, *Metro*, http://metro.co.uk/2016/06/23/ouch-people-have-lost-thousands-by-throwing-out-these-valuable-items-5962268/.

"From Trash to Treasure: 7 of the Most Valuable Items Ever Found in a Dumpster." Hometown Dumpster Rental, https://www.hometowndumpsterrental.com/blog/from-trash-to-treasure-7-of-the-most-valuable-items-ever-found-in-the-dumpster.

Grundhauser, Eric. "The Treasures in the Trash Collection." *Atlas Obscura*, https://www.atlasobscura.com/places/the-treasures-in-the-trash-collection.

Hanlon, Mike. "The World's Most Valuable Barn Find: 60 Rare Cars Untouched for 50 Years." *New Atlas*, December 14, 2014, https://newatlas.com/worlds-most-valuable-barnfind/35178/.

"How Big Are the Collectible Markets? Are We Really Spending $200 Billion Every Year on Them?" Hobby db, April 16, 2016, https://blog.hobbydb.com/2016/04/16/how-big-are-the-collectible-markets/.

"How Much Is Your Stuff Worth?" *Consumer Reports*, October 2012, https://www.consumerreports.org/cro/magazine/2012/10/how-much-is-your-stuff-worth/index.htm.

"Look What I Found! 5 Treasures Found among Trash." *A Whole Lot of Nothing* (blog), January 27, 2010, https://allthingsmundane.wordpress.com/2010/01/27/look-what-i-found-5-treasures-found-among-trash/.

Miller, Jim T. "How Much Is Your Old Stuff Worth?" *Huffington Post*, May 24, 2014, https://www.huffingtonpost.com/jim-t-miller/how-much-your-old-stuff-i_b_5384181.html.

Mulroy, Zahra. "9 of the Most Expensive Things People Have Accidentally Thrown Away Before Realizing Their Huge Mistake." *Mirror*, June 2016, https://www.mirror.co.uk/money/9-most-expensive-things-people-8243526.

Reif, Rita. "Declaration of Independence Found in a $4 Picture Frame." *New York Times*, April 3, 1991, https://www.nytimes.com/1991/04/03/arts/declaration-of-independence-found-in-a-4-picture-frame.html.

Singh, Kyli. "The 10 Most Valuable '*Antiques Roadshow*' Items." Mashable, https://mashable.com/2014/08/28/antiques-roadshow-valuable-items/#TgJtUhgfdgqs.

Vaccaro, Amie. "10 Tips from an Eco-Capitalist: Tom Szaky Shares TerraCycle's Secrets to Success." *Green Biz*, March 24, 2009, March 24, 2009, https://www.greenbiz.com/blog/2009/03/24/10-tips-eco-capitalist-tom-szaky-shares-terracycles-secrets-success.

INDEX

ABOUT THE AUTHOR

Jeff Dondero has a diverse background and experience in writing, ranging from web content, B2B, books, hard news, and interviews to feature writing, to internet content. After studying Business and Law at the University of San Francisco, and Communications, Broadcast Arts, English, and Journalism, at the San Francisco State University, he began his career as stringer and freelancer for the *San Francisco Examiner*. He worked as a reporter, editor, and contributor for several newspapers and magazines in the San Francisco bay area, was the entertainment editor for *The Marin Independent Journal*, wrote for KTVU-TV, toiled in a trade magazine mill, and co-created a website dedicated to the technology of sustainable construction industries, the *Green Building Digest*.

He was invited as a writer-in-residence at the art colony in Rancho Vista, Arizona, in 2014, where he wrote a slim volume of poetry and researched material for upcoming books. He continues to expand his national readership with books, social media, various writers' blogs, websites, and radio and television appearances.

His books include: *The Energy Wise Home*, *The Energy Wise Workplace*, *Throwaway Nation*, and *Super Cities* (to be released in 2019) published by Rowman & Littlefield. Other books include: *So Do You Want to Survive A Natural Disaster*, *The Marin Companion* (to be released in 2019), and *Brutal Beauty*, a collection of poems written as a writer in residence in Linda Vista art colony.

Mr. Dondero lives and works in Rohnert Park, Sonoma County, California.